Computational Molecular Bioscience: Concepts

Computational Molecular Bioscience: Concepts

U. S. Raghavender, Ph.D.

Delve
PUBLISHING

www.delvepublishing.com

Computational Molecular Bioscience: Concepts

U. S. Raghavender, Ph.D.

Delve Publishing
2010 Winston Park Drive,
2nd Floor
Oakville, ON L6H 5R7
Canada
www.delvepublishing.com
Tel: 001-289-291-7705
001-905-616-2116
Fax: 001-289-291-7601
Email: orders@arclereducation.com

ISBN: 978-1-77361-242-3 (Hardcover)

ABOUT THE AUTHOR

U. S. Raghavender, Ph.D.

Dr. U. S. Raghavender did his Masters in Physics, with specialization in Condensed Matter Physics, followed by his doctoral thesis in X-ray Crystallography of Designed molecules. This was followed by post-doctoral stints in India and abroad, in both academic institutions and non-academic institutions. He has been working on computational approaches of biomolecules, since the past 12 years. He has authored book chapters, published research in internationally acclaimed journals. He also serves as a peer reviewer of research articles published in both national (*Journal of Biosciences*) and international journals (*BMC Bioinformatics*, *Nucleic Acids Research etc.*). He has a long term goal of addressing computational aspects of regulatory mechanisms in cell.

TABLE OF CONTENTS

PREFACE

This preface provides information that I would recommend for someone reading and using the book. This book is written keeping in mind the undergraduates and graduates studying life sciences in universities. This could serve as a supplement for people involved in research.

With the flood of genomics and proteomics data, it is imperative that students in biology have a basic understanding of the data in biological repositories. Academic institutions and focussed organizations in genomics and proteomic research, in today's world, share both raw and processed data in dedicated servers. These datasets are accessible, for free, to anybody on this planet. Understanding how these datasets were produced requires one to study the research which prompted their existence (for example PDB or UniProt). This comprehension can result in a greater appreciation of the resource. We have provided as much information as is needed to jumpstart a curious reader's comprehension of the actual data from the repositories using computers.

The information processing done in biology using computation (algorithms) constitutes what is known as **Computational Biology**. The algorithms could be graded from simple to complex, based on the assumptions employed. In this book, we present basic concepts in computational molecular biology in the form of simple capsules. What I mean by that is the reader, with a laptop or a computer, can immediately start executing the code in the book to understand the concepts at a deeper level. As the focus is not on algorithms, we have kept the language quite simple. The places which demand detailed explanation have been dealt with accordingly. Otherwise, the presentation has been kept short. Basically, we start from sequences, then go to structures and try to find an evolutionary connection between sequences and structures.

1. We start with an Introduction to biology and genes and proteins.
2. Following which, we introduce the basics of **Python**. This is a must for any curious student starting in computational biology.
3. In the third chapter, we introduce sequences in biology and the commonly used formats (e.g., FASTA). We also introduce different ways of comparing sequences. BLAST and EMBOSS suites are introduced.
4. R, the statistical language, is introduced. This is of utmost importance, given the fact that any biological observation and experimental results

seek statistical significance. The basics are presented along with the processing of FASTA files.

5. The next chapter is on *sequence statistics*. With the computational background set for statistical analysis, we embark on common sequence statistics.

6. Researchers work with a large number of sequences. Hence, comparing them through *multiple sequence alignment* and the different approaches are presented in the next chapter.

7. With the flood of genomics, the set of sequences produced by large and computational annotation of gene function(s) becomes imperative. We introduce the KEGG database and present recipes for accessing the data from the resource using computation.

8. A large number of DNA sequences, upon translation, do fold into 3D structures. We introduce the reader to the basics of structural biology. We also show how one can handle structural data using computational techniques.

9. Lastly, we introduce evolution. Wherein, we try to connect sequences with structures.

10. Two appendices are provided. These are meant for the reader to explore interesting topics in programming like regular expressions and data analysis using *pandas*. But, these also provide not only a quick introduction but also a step forward to further explore the concepts using these topics.

Every chapter begins with a small introduction followed by explanation of concepts and a simple example. Computational Code is given in bounded box, and in a font (*Centaur*) different from the rest of the text (*Times New Roman*). References are provided at the end of each chapter, as needed. Each chapter is written to be self contained. I would also suggest that the reader become familiar with **IPython** or **Jupyter** notebooks and **RStudio**. Most of the code in this book has been tested on these notebooks and should be reproducible.

Finally, I hope that this book will help readers use computers and or laptops productively in their studies in the field of computational biosciences.

Author

1

INTRODUCTION

Biology is broadly defined as the study of living beings. Biologists study living things by performing experiments, collecting and interpreting data. With the advent of technology in the biological disciplines, we have a lot of experimental data generated within a short time. The sophistication of the experimental procedures and the vast volume of data, requires a concomitant match in exploration using computation. The most commonly cited example for this dichotomy is the data generated out of genomics experiments. At present, we can generate a large volume of DNA sequence data in a short space of time. We will be left with a large sea of adenosines (A), thymines (T), cytosine (C) and guanines (G). Computational scientists seek answers to many questions, among which a few common ones are given below:

- Which part(s) of DNA control the various chemical processes of life?
- We know the structure and function of a few proteins, then how can we predict the function of new proteins?
- How do the structures of unsolved proteins/DNA/RNA look like?

Bioinformatics is a science to understand biology using computational techniques. Essentially, it is a collection of tools which can be used to explore and answer a few particular questions in biology. It can be put under the umbrella of a much wider field of *Computational Biology*. Computational Biology aims at exploring large amount of biological data with quantitative analytical techniques. We will try to explore a few of these questions in the present book, with a mix of Bioinformatics tools and a bit of computation. The distinction of the two terms is irrelevant to experimental molecular biologists, but for scientists actively engaged in the fields the connotations are different. We will avoid making this distinction in this book.

The field of Bioinformatics gained attention with the painstaking efforts of a few dedicated scientists who had expertise in the areas of statistics and pattern recognition. Presently, the field welcomes people from varied backgrounds including mathematics, statistics, computer science, linguistics, chemistry, and biology. Naturally, each and every tool has a pitfall(s) in bioinformatics. Users need to have a good layman understanding of what each tool does and in what contexts it can be used to answer specific biological questions. There is no need to overemphasize the importance of bioinformatics and computational science in the era of *-omics*. With a firm grasp of core concepts of few basic and commonly used tools, one can jump start into the wide field of biology using computation.

BIOLOGY AND COMPUTATION

The genetic code of an organism is encoded in its DNA. DNA has coding regions which result in the production of the *workhorses* of a cell - *proteins*. Both DNA and proteins are linear polymers of fundamental building blocks. DNA is composed of four bases adenosine (A) - thymine(T) - cytosine(C) -guanosine(G). Proteins are made of 20 amino acid residues. The chemical and molecular structures of the building blocks of DNA and proteins are shown in Figure 1.

Both the DNA and proteins are linear chains of these fundamental building blocks. The linear chains are represented, in shorthand, as one-letter words in 1- dimension. These 1D sequences have a definite

3-dimensional (3D) structures. In a protein, amino acids are connected with each other through peptide bonds. Hence, proteins are also known as polypeptide chains. We will discuss the geometry of polypeptide chains in the following chapters. DNA, unlike protein, is *double-stranded*. For simplicity, we will try to understand a single strand. It has a backbone consisting of repetitions of a sugar molecule called 2'-deoxyribose attached to a phosphate group. The sugar molecule has 5 carbon atoms. The bond that results in the formation of DNA backbone is between the 3' carbon of one unit, the phosphate residue, and the 5' carbon of the next unit. This gives DNA stand an orientation, which by convention, starts at the 5' end and finishes at the 3' end (5' ® 3'). Bases are attached to the 1' carbon of the backbone. A nucleotide is the basic unit of DNA comprising of a sugar, phosphate, and a base. Although bases and nucleotides are not the same thing, we can speak of DNA having 200 bases or nucleotides. We know from the work of Watson and Crick that DNA is double stranded and the two strands are held together in a fixed fashion. In this model, A always pairs with T and C pairs with G by making hydrogen bonds. Bases A and T are said to be complementary bases. Hence, given the sequence of one strand the sequence of the second strand can be guessed easily. We will not be considering RNA in the present book.

Gene sequences of related species of plants, animals and/or microorganisms show complex similarity patterns and their computational study constitutes one of the most fascinating aspects of studying evolution. Biological sequence analysis is one of the most common methods used to study sequence evolution and forms an important sub-discipline in computational biology. BLAST (Basic Alignment Local Search Tool) is the most commonly used tool, from National Center for Biotechnology Information (NCBI), for searching similar sequences. On a daily basis, every student and researcher, visits the NCBI's BLAST database for performing sequence based searches. Basically, it compares two sequences by trying to align them. It is also used to lookup sequences in a database. The algorithm starts by looking for exact matches, then expands the aligned regions by allowing for mismatches. We will be exploring some of these concepts in detail in following chapters.

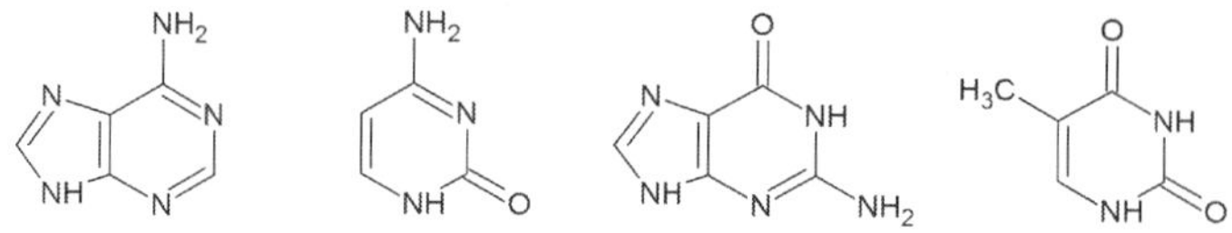

(i) Adenine (ii) Cytosine (iii) Guanine (iv) Thymine

Basic building blocks of Deoxy-ribose Nucleic Acid (DNA)

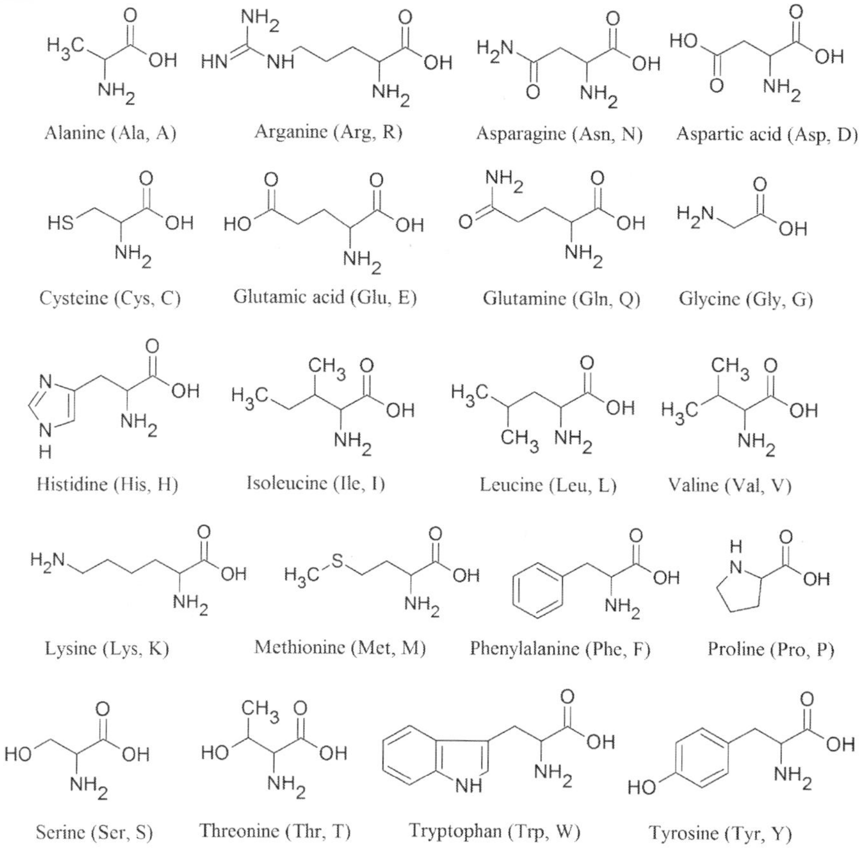

Basic building blocks of Proteins - 20 amino acids

Figure 1.1: Representation of the basic chemical structures of bases in DNA and amino acid residues in proteins.

INFORMATION AGE IN BIOLOGY

Biologists have been collecting, organizing, and interpreting a lot of experimental information for the past few centuries. The field of taxonomy in botany and zoology, is one good example of how the information revolution started in biology. It is very difficult, even for some gifted with very good memory, to memorize the complete taxonomy of a given species of plant or animal.

At the end of the 1800s, *vitalism*—the idea that life had features that were distinct from the rest of the physical world—was still popular. It was the parallel progress in genetics and biochemistry that helped bring *vitalism* to a close. It took until the 1940s for the two fields to overlap, as researchers started to study the genetics of biochemical pathways and the biochemistry of heritable material. This is when the central role of DNA became clear. And that's what inspired Watson and Crick (plus a number of others) to think that understanding the structure of DNA would be critical.

The double helix moved biology into the information age, as Lander put it, since it answered some key questions: how can you store information, faithfully replicate it, but still allow room for the new variants that drive evolution? The identification of the genetic code was the next (and obvious) step in understanding how information was stored in a string of nucleotides. The problem, Lander said, was that you couldn't actually read any of this—at least not as it exists in a cell.

From a biochemical perspective, all DNA was more or less the same, so you couldn't deal with that information by the standard methodology of the time. It took the development of molecular biology and recombinant DNA to start dealing with the actual information content.

"How do you ask how to read DNA?" Lander asked. "You ask the master—the cell. The cell is in the business of reading the information in DNA." Molecular biology and biotechnology developed around the purification and use of the proteins used by the cells themselves to manipulate DNA.

The next hurdle, according to Lander, was that we had no idea how to apply the principles of molecular biology to something as big as the human genome in a systematic way. That problem first started to crack with the development of RFLP mapping, when human genes were first associated with a nearby DNA feature that could only be revealed using the techniques of molecular biology. In principle, these worked. But Lander said without the genome sequence, each research group had to essentially start from scratch. For an in-depth description of RFLP mapping, reader is referred to a book hosted at NCBI named Modern Genetic Analysis.

Conceptually, he said the key step was the development of a hierarchical map. Lay out genetic markers on a map, identify the DNA associated with those markers, and then dig down into the actual DNA sequences. The first human genetic map appeared in 1987. That set the stage for the genome sequencing to kick off in earnest in the 1990s. The final draft was announced in 2003, on the 50th anniversary of the Watson and Crick paper.

Simplified representation is the primary step in comprehending the hidden complexities in genetic code and functioning of proteins. Exploiting the tremendous power of computation, demands that the these building blocks be represented as a single alphabet (Figure 1). This has both merits and demerits. The single letter representation takes away all the feature (chemical, physical and biological), but is the preferred way of representing sequences. Hence, we will be exploring ways of sequence manipulation in this book.

We will explore the concept of information encoding in our DNA in the next section.

DNA

DNA stands for *deoxyribonucleic acid.* Like a protein, a molecule of DNA is a chain of simpler molecules. Actually it is a *double* chain, but let us first understand the structure of one simple chain, called strand. It has a backbone consisting of repetitions of the same basic unit. This unit is formed by a sugar molecule called 2'-deoxyribose attached to a phosphate residue. The sugar molecule contains five carbon atoms, and they are labelled 1' through 5' (Figure 2). The bond that creates the backbone is between the 3' carbon of one unit, the phosphate residue, and the 5' carbon of the next unit. For this reason, DNA molecules also have an orientation. This by convention starts at 5' end and finishes at 3' end.

Attached to each 1' carbon in the backbone are other molecules called bases. There are four kinds of bases: adenine (A), guanine (G), cytosine (C), and thymine (T). Bases A and G belong to a group of substances called *purines,* whereas C and T belong to the *pyrimidines.* When we see the basic unit of a DNA molecule as consisting of the sugar, the phosphate, and its base, we call that a nucleotide. Thus, although bases and nucleotides are not the same thing, we can speak of a DNA molecule having 200 bases or 200 nucleotides. A DNA molecule having a few nucleotides is referred to as an *oligonucleotide.* DNA molecules in nature are very long, much longer than proteins. In a human cell, DNA molecules have hundreds of millions of nucleotides.

DNA is double stranded. The two strands are held together by complementary base pairing. This was the original contribution of Watson and Crick. In this scheme, A always pairs with T and C with G. This is also known as *Watson-Crick base pairing.*

This base pairing leads to the anti-parallel strands making DNA combining to form a double helix with bases on the inside of helix and sugar-phosphate forming the backbone of the helix.

Now let us see how information is encoded in DNA.

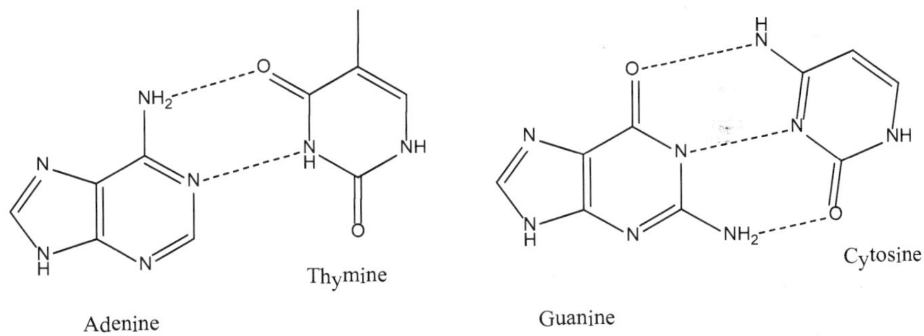

Figure 1.2 Watson-Crick Base pairing illustrated.

The importance of DNA molecules is that the information necessary to build each protein or RNA found in an organism is encoded in DNA molecules. For this reason, DNA is sometimes referred to as "*the blueprint of life.*" Each cell of an organism has a few very long DNA molecules. Each such molecule is called a chromosome.

The first important thing to know about this DNA is that certain *contiguous stretches* along it encode information for building proteins, but others do not.

The second important thing is that to *each different kind of protein in an organism there usually corresponds one and only one contiguous stretch* along the DNA. This stretch is known as a gene.

THE CENTRAL DOGMA

Every organism possesses a unique set of traits that distinguish it from members of other species and, in subtler ways, from other members of its own species. These traits, which are passed down from the organism's ancestors, are complex manifestations of units of heritable information known as genes.

How does an organism store, transmit, and interpret this genetic information that specifies its identity?

The central dogma of molecular biology posits that cells possess chemical codes detailing the assembly instructions for macromolecular machinery that promotes reactions underlying the fundamental physiological processes and states characteristic of the cell. Stored genetic material, in the form of DNA, is decoded to produce RNA intermediates, which are in turn decoded to build instruments of biological activity, proteins

DNA ® RNA ® Protein

RNAs and proteins that are created in this way are referred to as gene products, and synthesis of such structures is known as gene expression. The molecular structures and reactions underlying this flow of information are currently understood in some detail.

GENETIC INFORMATION AT THE MOLECULAR SCALE

On a molecular level, a gene is a sequence of deoxyribonucleic acid, or DNA. A molecule of DNA consists of two unbranched chains coiled around each other to form a double helix. Each strand is built of nucleotide subunits, which are each in turn comprised of a five-carbon sugar (the eponymous deoxyribose) and a phosphate group–the two components that form the backbone of a DNA strand–as well as a nitrogenous base with any of four possible chemical compositions. The four bases define nucleotides known as adenine (A), guanine (G), thymine (T), and cytosine (C), and it is the identities of these nucleotides, arrayed in various combinations, that encode genetic information.

The double-stranded nature of DNA adds redundancy, as particular base pairs (bp) co-occur at the same position across sister strands and share weak bonds. These specifically pair A with T and G with C. The sequences of the two strands are thus complementary, with each reflecting the content of the other in a set of corresponding bases. This repetitive nature allows the fidelity of the sequence to be preserved in the event that one strand is damaged.

The information contained on each strand is also directional, as linkages within a DNA molecule imbue the structure with chemical polarity. Within the chain, a nucleotide is bound to the nucleotide that precedes it through its own phosphate group and to the nucleotide that follows it at its deoxyribose group. Under a particular convention for numbering the carbons within the five-carbon sugar, these linkages occur at the fifth carbon (which bears the phosphate group) and third carbon, respectively. Additional nucleotides can only be added to the end of the strand at which a nucleotide possesses an accessible third carbon. All reactions relying on the synthesis of new nucleic acid polymers–a category which includes both DNA replication and DNA sequence decoding (*to be explained*)– must thus proceed along the polymer in a single direction known as 5' to 3' *("five-prime to three-prime")*. This is the sole direction in which a strand "runs" and in which its genetic information can be read.

The two strands within a double helix are oriented such that the 5' end of one strand is paired with the 3' end of its sister. As the two strands run in opposite, or antiparallel directions, each sequence is, relative to its sister, a series of opposite bases running backwards, or the reverse complement.

Figure 1.3: Double-stranded DNA.

Genetic information is extracted from DNA by transcription, in which the double helix unfurls and one strand serves as the template for the production of a molecule of ribonucleic acid, or RNA from the gene source. The structure of RNA is similar to that of DNA, but its phosphate-sugar backbone incorporates a slightly different sugar (ribose) and the base uracil (U) is substituted for thymine. Like thymine, uracil associates with adenine bases, although RNA is typically single-stranded and exhibits base pairing primarily in complex structures in which a single RNA strand partially folds against itself.

Transcription is capable of generating RNA products, known as transcripts, using either strand within a double helix as a template. The protein that catalyzes transcription, RNA polymerase unwinds a small portion of the double helix, which allows free RNA nucleotides to base pair to the template. The polymerase then links nucleotides into a continuous polymer by spurring the formation of phosphodiester bonds between the ribose sugars and phosphate groups of successive nucleotides. RNA, like DNA, can be synthesized solely from its 5' end along its 3' end, and this is the direction in which RNA polymerase moves down a nascent transcript.

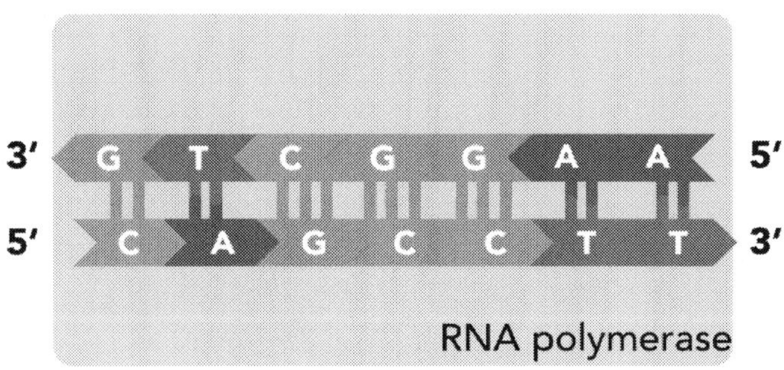

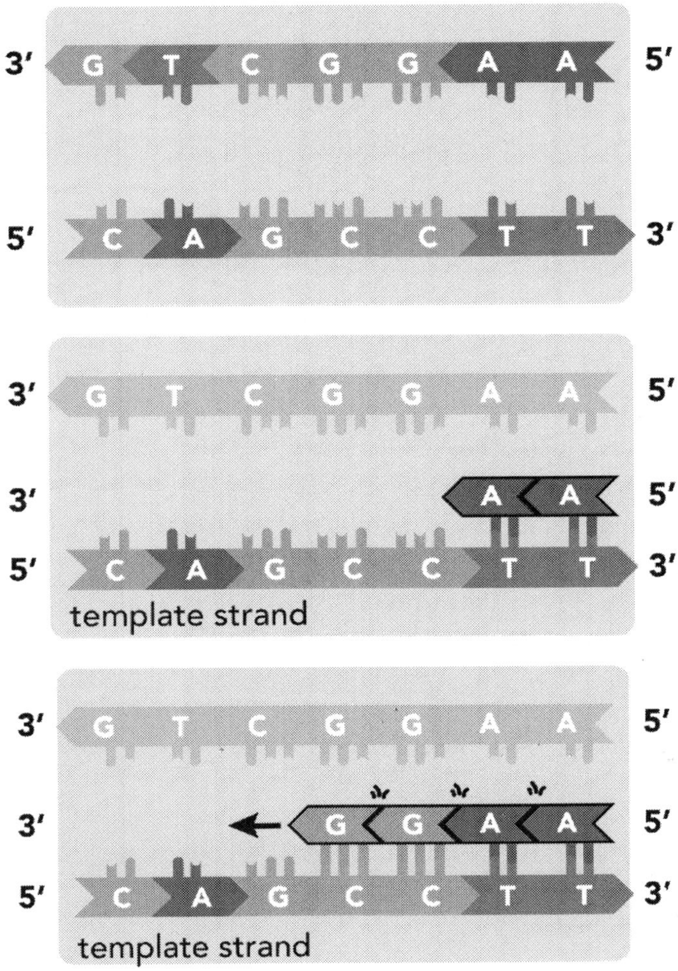

Figure 1.4: Mechanism of transcription. RNA polymerase (depicted as a yellowish blob surrounding the DNA).

As the RNA lies antiparallel to its DNA source material, the sequence of a transcript will always be the reverse complement of its template. The template's sister strand, also the reverse complement of the template, features the sequence that will actually match the transcript (with U-T substitutions). This strand is accordingly known as the coding strand. Which strand is coding and which is template can vary among the many genes present within a stretch of DNA.

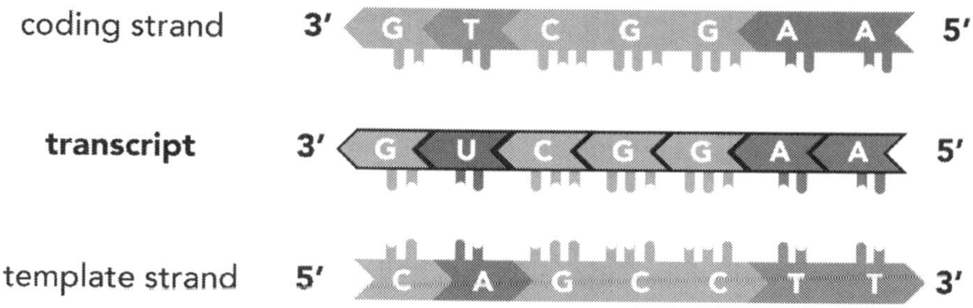

Figure 1.5: RNA transcript is reverse complement of its template DNA.

Genes–sequences that encode transcripts in this way–constitute only a fraction of an organism's total DNA content.

How, then, is a particular DNA sequence recognized as a gene, and how is it targeted for transcription?

A sequence that serves as template for an RNA molecule is prefaced by a promoter region, a stretch of DNA that defines the gene's start site. The promoter serves at the binding site at which RNA polymerase will be loaded onto the sequence prior to transcription.

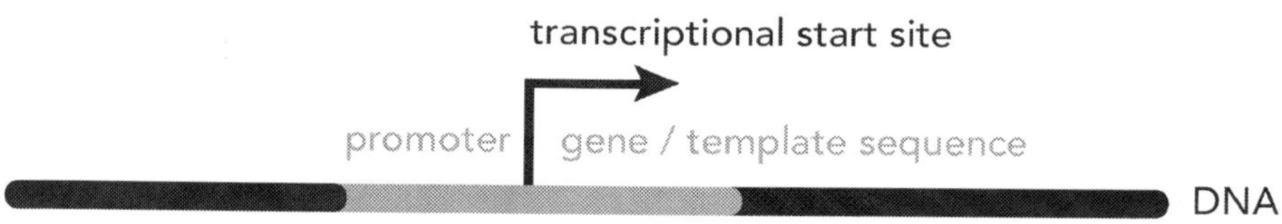

Figure 1.6: Transcription.

Only a subset of RNA molecules provide template from which proteins are produced. These are messenger RNAs (mRNAs). Such RNAs undergo a series of post-transcriptional modifications that mark them as information sources for future proteins.

In eukaryotes (cells that contain a defined nucleus in which genetic information is stored, among other membrane-delineated compartments that segregate biochemical reactions), a pre-mRNA, or primary transcript, the native RNA form that is the direct product of transcription, is converted into a mature mRNA that is ready to be decoded into protein through three principal processing steps: 5' capping, polyadenylation, and RNA splicing.

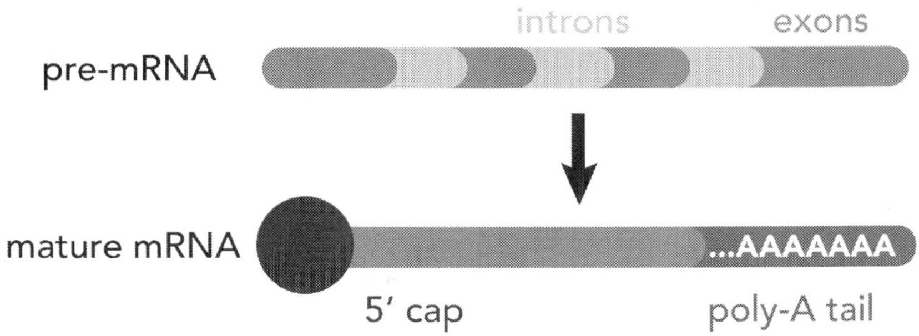

Figure 1.7: RNA processing

First, even as the RNA is still being transcribed, the 5' end of the nascent transcript is fitted with a 5' cap, a single modified guanine nucleotide. This cap both identifies the transcript as an mRNA (for purposes of nuclear export and protein production) and is protective against degradation.

After translation is complete, the 3' end of the transcript is extended with the addition of roughly 200 nucleotides, almost all of which are adenine bases. This poly-A tail is not specified with the DNA content of the gene and is synthesized without a template. Like the 5' cap, this modification, polyadenylation, marks the RNA as protein-coding and increases the stability of the molecule.

Not all regions within a coding transcript encode for protein sequence. In the third stage of RNA processing, noncoding intron regions are excised from mRNA precursors, leaving only coding regions, or exons, comprising mature mRNA. This editing step, RNA splicing, occurs by a complicated mechanism in which a large set of proteins and one class of RNAs that are not protein-coding in themselves, small nuclear RNAs (snRNAs) constitute a spliceosome complex that catalyzes the formation of loops within intron sequences followed by their complete cleavage from their adjacent exons and the connection of free exon ends. In some transcripts with numerous exons, splicing reactions may combine subsets of exons to produce distinct combinations and thereby distinct protein-coding sequences from a single primary transcript sequence. This flexibility in RNA maturation is known as alternative splicing.

Once mRNAs have been processed, genetic information can then be extracted from RNA by translation, in which an RNA transcript serves as the template for the production of a protein, which consists of one or more polypeptide chains. A polypeptide chain is a linear sequence of amino acid subunits. Twenty different amino acids can appear in polypeptides, yet they are specified from a nucleotide code comprised of just four distinct base types. This level of variation is possible by specifying amino acids with groups of three consecutive nucleotides. The set correspondence between RNA triplets, or codons, and their resulting amino acids is the genetic code.

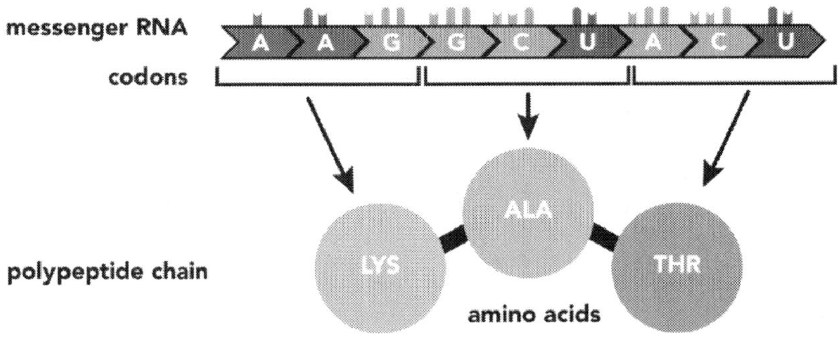

Figure 1.8: Translation

Proteins are essentially gene products. A protein, as we have seen, is a chain of amino acids. Therefore, to "specify" a protein all you have to do is to specify each amino acid it contains. And that is precisely what the DNA in a gene does, using triplets of nucleotides to specify each amino acid. Each nucleotide triplet is called a codon. The table that gives the correspondence between each possible triplet and each amino acid is the so-called genetic code. See the diagrams below for understanding which triplets code for which one of the naturally occurring amino acids. The *start codon* AUG codes for Met, and the *stop codons* are UAA, UAG and UGA.

Figure 1.9: An example of mRNA to protein translation.

In this book we will not be focussing on the basics of biology, for which the reader is referred to standard textbooks in the area. Rather, we fill shift our focus on how to use computation to get meaningful insights into biology.

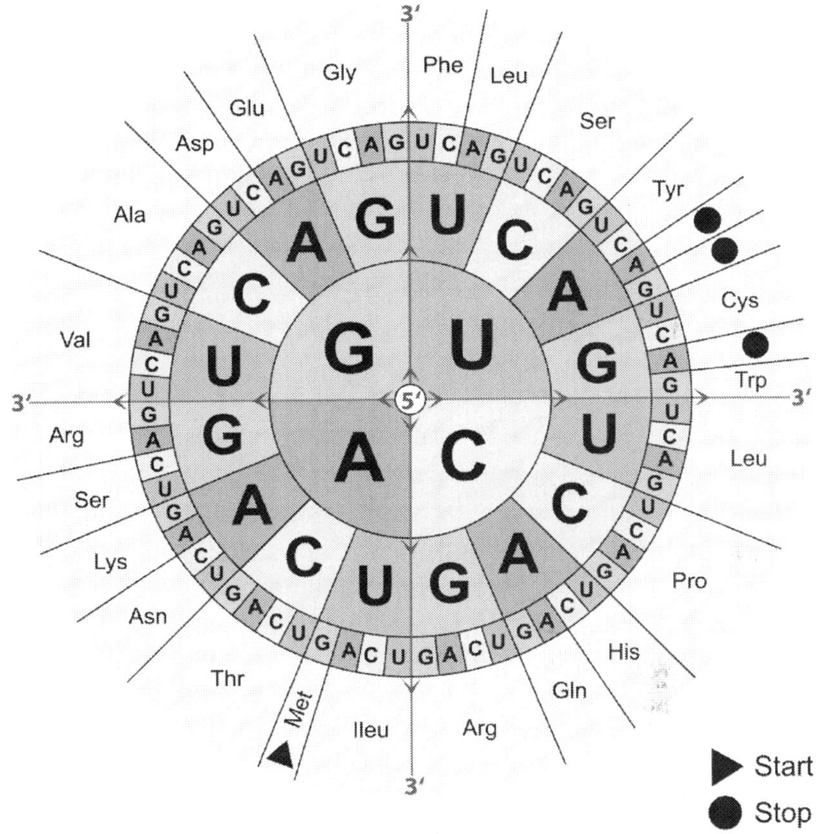

Figure 1.10: Standard Genetic Code.

HUMAN GENOME PROJECT (HGP)

Any book on computational biology is incomplete without due reverence to the multinational efforts to decipher the genetic code of *Homo sapiens*. Draft genome was published by HGP in 2001 and the complete genome in 2003. The main findings of the above studies were:

- Approximately 22000 protein-coding genes in humans.
- Human genome has substantial segmental duplications (repeated sections of DNA).

The genome paved the way for intensive studies on many aspects of human biology. Ranging from understanding gene/protein function to disease implications to evolution, almost every field in biology came under the grip of human genome. The project also inspired genomic works in other areas as well, like plants and agriculture. It is now possible to analyze the totality of an organism's DNA content, its genome; its complete set of mRNAs, its transcriptome; and every protein that it contains, its proteome.

MOLECULAR MODELLING

At almost the same time as the publishing of human genome, there was a remarkable rise in the interest in performing molecular modelling and simulation. Scientists have agreed upon one thing, and it is that biological systems are highly complex. A deterministic approach to understanding biology is that - life is a product of 'upward' causation from genes to cells, organs, systems and whole organism. Then, the question arises where should one focus for modelling? In other words, what should be the starting point. The answer to the question lies in your understanding of the problem in the light of resources accessible to you and the level of expertise in the subject area. Computationally all these can be addressed, at least qualitatively.

It is worthwhile to keep in mind the hunch of Denis Noble - "*My hunch is that twenty-first century will witness the consummation of the marriage of mathematics and medicine. Computational biology will then simply have been a stepping stone to an even greater transformation of biological science.*"

TOOLS IN INFORMATION AGE

With the data being generated at a rapid pace daily in biology, demands tools which can be used to process the data and comprehend the information. Computers and high-level computer languages are the preferred tools for addressing the problem of information processing in biology and its subfields. We will be directing our focus on understanding the two important languages, Python and R, which have found wide usage in present day computational biology.

REFERENCES

1. https://arstechnica.com/science/2012/12/eric-lander-describes-how-biology-entered-the-information-age/
2. http://biobook.kuensting.org/bb/genetics/dna.htm
3. https://www.ncbi.nlm.nih.gov/books/NBK21320/
4. Venter, JC; et al. The sequence of the human genome. *Science*. 291: 1304–1351 (2001).
5. Venter D. A Part of the Human Genome Sequence. *Science*. 299: 1183–4 (2003).
6. Noble D. The Rise of Computational Biology. *Nat. Rev. Mol. Cell. Biol.* 3: 460-463 (2002).

2

PYTHON IN BIOINFORMATICS

Python is by all means a beautiful and simple computer language. In recent years, its popularity in bioinformatics community has increased substantially. Technically, Python is a powerful language with respect to computing paradigms. It has the skeleton of a *procedural* languages (like Prolog) with significant influence from *functional* languages (like LISP) and is a superb object-oriented language (like C++). We start this chapter by introducing Python with instructions for installing and running Python on your computer. Although, it is assumed that the reader is familiar with installing packages in different operating systems like Windows or OSX or Linux. We will explore different avenues in sequence and structural biology which can be addressed by scripting in Python in this introductory chapter.

Let us start by installing and running Python on your system.

Installing Python

Your system, for OSX or Linux-based operating systems, probably shipped with preinstalled Python. Else, one can download the latest release from

https://python.org/download/

For Linux distributions, installing through package mechanisms is suggested, whereas for OSX and Windows installers are available. It is highly recommended to install a self-contained distribution of scientific Python packages for computing like Anaconda (anaconda.org).

RUNNING PYTHON

You can start Python in one of the three ways:

- Type python2.x or python3 on the command line.
- Run an IDE. Python comes with one called IDLE.
- Run *Spyder*, an IDE which comes with *Anaconda* distribution.

For all the help, readers are requested to check the documentation which comes with the Python installation. Good tips and solutions can also be obtained at sites like *stackoverflow.com*.

For a good Bioinformatics background, I would suggest referring to the following two books:

- *Bioinformatics: Sequence and Genome Analysis*, by David W. Mount (Cold Spring Harbor Laboratory Press)
- *Bioinformatics and Functional Genomics*, by Jonathan Pevsner (Wiley-Blackwell)
- *Learning Python* - Powerful Object-Oriented Programming, by Mark Lutz (O'Reilly Media)

We begin this chapter with simple string manipulations, basic to any sequence analysis. It is assumed that the reader would familiarize themselves with the basics of Python. The book *Learning Python* is recommended for beginners.

STRING OPERATIONS

Let us start with some basic operations on strings. For most part in bioinformatics, you will be dealing with strings. There are four binary operators that act on strings:

- *in*
- *not in*

- +, and
- *

The first three have both operands as that of *type* string. The last operator (*) takes an integer as the other operator. Substrings are extracted using *subscription* (one-character) or *slicing* (longer).

```
>>> 'TATA' in 'ATTTATATTTTTTTTAATATA'
True
>>> 'AAA' in 'ATTTATATTTTTTTTAATATA'
False
>>> 'ATTA' not in 'ATTTATATTTTTTTTAATATA'
True
```

One can concatenate strings to produce a new string.

```
>>> 'AT' + 'CG'
'ATCG'
>>> 'ACD' + 'KLY'
'ACDKLY'
```

You can repeat a string a certain number of times by using * operator.

```
>>> 'AT' * 6
'ATATATATATAT'
```

One can extract a one-character string by subscription.

```
>>> 'ATTTATATTTTTTTAATATA'[0]
'A'
>>> 'ATTTATATTTTTTTAATATA'[1]
'T'
```

One is allowed to use negative indices; the last character is at index *-1*:

```
>>> 'ATTTATATTTTTTTAATATA'[-1]
'A'
>>> 'ATTTATATTTTTTTAATATA'[-10]
'T'
```

You can expect *IndexError* if the index is out of range.

With slicing, one can extract substrings consisting of a series of characters. Slicing is the most commonly used operation in Python and you'll be using it very often on strings.

A slice [p:q] is read as 'from character *p* up to but not including *q*'. Let us see it on a few examples:

```
>>> 'ATTTATATTTTTTTAATATA'[1:3]
'TTT'
>>> 'ATTTATATTTTTTTAATATA'[10:-1]
'TTTTTAATAT'
```

If both the indices are omitted, then the entire string is selected. If a third number is specified, following a colon, it indicates the number of characters to *skip* after each one that is included (known as *step*).

```
>>> 'ATTTATATTTTTTTTAATATA'[0:7:2]

'ATA'
```

There are some commonly used methods of the *str* type. Syntax of some are given below:

your_string.count(new_string[, start[, end]])

your_string.find(new_string[, start[, end]])

your_string.startswith(new_string[, start[, end]])

A few examples are given below (these are self-explanatory) :

```
>>> 'ATTTATATTTTTTTTAATATA'.count('TT')

5

>>> 'ATTTATATTTTTTTTAATATA'.find('TT')

1
```

returns the first location of match.

```
>>> 'ATTTATATTTTTTTTAATATA'.find('TT', 5, 14)

7

>>> 'ATTTATATTTTTTTTAATATA'.startswith('TT')

False

>>> 'ATTTATATTTTTTTTAATATA'.startswith('TT', 1)

True
```

Many of these built-in methods can be used for varied searches in sequences (for example, checking for recognition site etc.). Let us start with an example by *reverse transcribing a DNA* sequence. We will be writing it as a function, using the standard libraries of Python.

```
def revComp(seq, isDNA=True):
        from string import maketrans
        if isDNA:
                seq = seq.replace('U', 'T')
                translateTable = maketrans('ATGC', 'TACG')
        else:
                seq = seq.replace('T', 'U')
                translateTable = maketrans('AUGC', 'UACG')
        complement = seq.translate(translateTable)
        reverse_complement = complement[::-1]
return reverse_complement
```

This code relies on *translate* and *maketrans* functions imported from the string module; allowing a character substitution table to be used for equivalently positioned letters in two strings. The user-defined function *isDNA* also checks for wrong letters (U in place of T).

READING FILES

When reading an external file in Python, one needs to specify the *location* and *name* of the file. This is known as *specifying the path* to a file. We use python's open() function to get access to the contents of file. The open function passes a *file handle*, which is used to access and represent the file. For example,

```
path = 'example/sample.txt'

fo = open(path)
```

fo is known as file object. The function also takes an optional second argument. One can specify the *'mode'* in which the file should be opened.

Once you finish reading the file, you should close it using the close() function.

```
fo.close()
```

The file object allows to read the data in the file using functions:

read() - reads the entire file, into a single string;

readline() - one line from a file; each line ends with a newline ('\n') character;

readlines() - reads the entire lines into a list, with each line being an element in the list.

An alternative to manually closing a file in Python, is using with construct.

```
with open(path) as fo:

        for line in fo:

                # process line
```

READING FASTA FILES

FASTA is a text format to store nucleotide or protein sequences in one-letter codes. A FASTA file can have one or more sequence entries. Each entry has two parts.

- *Header*: every sequence entry starts with a header symbol '>'. It is followed by an identifier or a comment; usually gene or protein name with database codes are provided.

- *Sequence*: The next line is the sequence, till the next Header line is reached.

Header line is also known as comment line. An example of a FASTA sequence of a protein from the model plant *Arabidopsis thaliana* is given below:

```
>sp|O64587|ACD11_ARATH Accelerated cell death 11
OS=Arabidopsis thaliana GN=ACD11 PE=1 SV=1

MADSEADKPLRKISAAFKKLAIIVNSPNPEVPVTQFSHAC-
SLVSPLFGCLGIAFKFAEMD

YVAKVDDLVRASSSISTLVVMMDKDIEADCVRKAGSH-
TRNLLRVKRGLDMVKVLFEQIIA

SEGDNSLKDPATKSYAQVFAPHHGWAIRKAVSLGMYALP-
TRAHLLNMLKEDEAAAKIHMQ

SYVNSSAPLITYLDNLFLSKQLGIDW
```

Often one encounters many sequences, sometimes the whole genome, in a single FASTA file. We will write a function to read such a file, and generate a list of individual sequences.

```
def readFasta(name):
        fo = open(name)
        seqs = []
        seqFrags = []
        for line fo:
                if line.startswith('>'):
                        if seqFrags:
                                seq = ''.join(seqFrags)
                                seqs.append(seq)
                        seqFrags = []
                else:
                        seq = line.rstrip()
                        seqFrags.append(seq)
        if seqFrags:
                seq = ''.join(seqFrags)
                seqs.append(seq)
        fo.close()
return seqs
```

As you grow comfortable using Python and packages, you will start thinking about additional checks that need to be performed. For example, if the file is a genuine FASTA file etc. BioPython is a nice package to get acquainted with for performing FASTA checks and manipulations.

READING PDB FILES

PDB stands for *Protein Data Bank.* It is a repository of biomolecular structural information founded in 1970s. The data on molecular structure, describing the three-dimensional coordinates, is stored in text format. A quick look at a PDB file will reveal that it contains more information than just 3D coordinates of individual atoms making a biomolecule (protein/DNA/RNA/carbohydrates/lipids/small molecules etc.). Usually we deal with spatial coordinates, specified by the string ATOM at the start of line. The coordinates (x, y, z) appear in the column ranges (30-37, 38-45, 46-53). Let us take a simple case of *calculating the centroid* of a given structure. We will develop a simple code, which can be made accurate as needed. Again, for serious calculations well-tested functions in packages like *BioPython* and *ProDy* are recommended. A brief explanation of the code is given below.

We start by opening the file, read all the lines and close it. Next, we initialize variables which represent the number of atoms and the totals for (x, y, z) coordinates to 0. If a line starts with ATOM, coordinates are extracted.

```
def calc_Centroid(pdb):
        fo = open(pdb)
        num_atoms = 0
        x_tot = y_tot = z_tot = 0
        for line in fo:
                if line[:6] == 'ATOM  ':                // note the ex-
tra spaces after the ATOM string !
                        num_atoms += 1
                        x = float(line[30:38])
                        y = float(line[38:46])
                        z = float(line[46:54])
                        x_tot += x
                        y_tot += y
                        z_tot += z
                fo.close()
                if num_atoms == 0:
                        x_avg = y_avg = z_avg = 0
                else:
                        x_avg = x_tot/num_atoms
                        y_avg = y_tot/num_atoms
                        z_avg = z_tot/num_atoms
                return (num_atoms, x_avg, y_avg, z_avg)
```

WRITING FASTA FILES

Let us now venture into writing files. We start with writing a sample FASTA file. Assume you have a string of one-letter codes of a sequence of interest. We will write a function, which will generate a FASTA for the sequence of letters.

```
def write_FASTA(header, sequence, fasta):

        fo = open(fasta, 'w')                    // open a file object
'fo' in write ('w') mode

        fo.write('> %s\n' % header)

        for (n, code) in enumerate(sequence):

                if n > 0 and n % 60 == 0:        // check if 60 one-
letters in a line

                        fo.write('\n')           // add a new line

                fo.write(code)

        if n % 60 != 0:

                fo.write('\n')                   // add a new line at
the end

        fo.close()
```

You can try improving the code for writing a FASTA file with multiple sequences.

READING MULTIPLE FILES IN A DIRECTORY

Often you would be writing a generic script to be executed on certain files stored in a particular location on your computer. We do thin in Python using the library '*os*' and its sub-module '*os.path*'. Let us take a simple example wherein we are interested in finding all files ending in suffix '*.pdb*' in a particular directory.

```
import os

def findPDB(dir, suffix):

        files = []

        dir_files = os.listdir(dir)

        for file in dir_files:

                full_path = os.path.join(dir, dir_file)

                // check if the above path is a directory. If yes,
recurse into it.

                if os.path.isdir(full_path):

                        files.extend(findPDB(full_path, suffix))

                elif file.endswith(suffix):

                        files.append(full_path)

        return files
```

RNA CODON TRANSLATION: A DICTIONARY EXAMPLE

Tabular data in Python is usually represented as a dictionary. Dictionaries are very important when handling voluminous structured data. Information from a row of database table can be thought of as (*key, value*) pair, with table's *key* acting as dictionary *key* and *value* is represented as a *tuple* of each column of data. Let us take a simple example of generating a dictionary for the RNA codon-to-amino acid translation table.

Table 2.1 RNA codon translation table.

		Second Base			
First Base	U	C	A	G	Third Base
U	Phe	Ser	Tyr	Cys	U
	Phe	Ser	Tyr	Cys	C
	Leu	Ser	stop	Stop	A
	Leu	Ser	Stop	Trp	G
C	Leu	Pro	His	Arg	U
	Leu	Pro	His	Arg	C
	Leu	Pro	Gln	Arg	A
	Leu	Pro	Gln	Arg	G
A	Ile	Thr	Asn	Ser	U
	Ile	Thr	Asn	Ser	C
	Ile	Thr	Lys	Arg	A
	Met	Thr	Lys	Arg	G
G	Val	Ala	Asp	Gly	U
	Val	Ala	Asp	Gly	C
	Val	Ala	Asp	Gly	A
	Val	Ala	Asp	Gly	G

We will attempt to encode this information into a Python dictionary.

```
codon_table = {
# U
'UUU': 'Phe', 'UCU': 'Ser', 'UAU': 'Tyr', 'UGU': 'Cys', # UxU
'UUC': 'Phe', 'UCC': 'Ser', 'UAC': 'Tyr', 'UGC': 'Cys', # UxC
'UUA': 'Leu', 'UCA': 'Ser', 'UAA': '---', 'UGA': '---', # UxA
'UUG': 'Leu', 'UCG': 'Ser', 'UAG': '---', 'UGG': 'Urp', # UxG
# C
'CUU': 'Leu', 'CCU': 'Pro', 'CAU': 'His', 'CGU': 'Arg', # CxU
'CUC': 'Leu', 'CCC': 'Pro', 'CAC': 'His', 'CGC': 'Arg', # CxC
'CUA': 'Leu', 'CCA': 'Pro', 'CAA': 'Gln', 'CGA': 'Arg', # CxA
'CUG': 'Leu', 'CCG': 'Pro', 'CAG': 'Gln', 'CGG': 'Arg', # CxG
# A
'AUU': 'Ile', 'ACU': 'Thr', 'AAU': 'Asn', 'AGU': 'Ser', # AxU
'AUC': 'Ile', 'ACC': 'Thr', 'AAC': 'Asn', 'AGC': 'Ser', # AxC
'AUA': 'Ile', 'ACA': 'Thr', 'AAA': 'Lys', 'AGA': 'Arg', # AxA
'AUG': 'Met', 'ACG': 'Thr', 'AAG': 'Lys', 'AGG': 'Arg', # AxG
# G
'GUU': 'Val', 'GCU': 'Ala', 'GAU': 'Asp', 'GGU': 'Gly', # GxU
'GUC': 'Val', 'GCC': 'Ala', 'GAC': 'Asp', 'GGC': 'Gly', # GxC
'GUA': 'Val', 'GCA': 'Ala', 'GAA': 'Glu', 'GGA': 'Gly', # GxA
'GUG': 'Val', 'GCG': 'Ala', 'GAG': 'Glu', 'GGG': 'Gly' # GxG
}
```

To view the dictionaries, it is suggested to use the module *pprint* as follows:

```
from pprint  import pprint as pp

>> pp(codon_table)
```

The way to access the information in a dictionary, is by providing 'key'; membership operators in and not in check for keys presence in the dictionary; max and min compare keys and all and any test keys. The reader should refer to Python's documentation on understanding the dictionary methods.

ANONYMOUS FUNCTIONS

Python provides a mechanism for defining functions without using *def*. Since they are not given any names, they are called anonymous. They are defined as an *expression* and not as a statement. This is known as calling a *lambda* expression. The syntax is as follows:

lambda args: expression (which uses *args*)

This definition can take one or more arguments, and one can use this anywhere in the code as a named function. An example:

```
>>> square = lambda x: x^2

>>> print square(2)            // Answer is 4
```

REGULAR EXPRESSIONS

Regular expressions are strings. These strings are used as *patterns* to perform match and search operations on *target* strings. A successful match operation yields a *match object*, which has quite a few fields and methods providing details about the match, such as the position within the target where the pattern was found. If the match or search fails, the result is None, rather than a match object. Python does distinguish between *matching* and *searching*. I leave it to the inquisitive reader to delve into technicalities. We will take a simple example to illustrate the concepts involved.

Assume you are working with multiple sequences contained in a FASTA file. We will see how we can extract entries using regular expressions.

Say, you want to extract the description line or header. Then the following will do it.

```
re.search(r'^>[^>]*', src).group()
```

What it means is 'match the following > which starts a line with all characters that aren't a >'.

As an aside, this is equivalent to *grep* operation on a Linux command line.

We can improve our approach by capturing two groups, namely *header* and *sequence* parts.

```
re.search(r'^>(.*)$([^>]+)', src, re.M).groups()
```

I leave it to the reader to derive meaning of this regular expression.

Please refer to Appendix 1 for a collection of useful regular expressions.

The following books are suggested for the interested reader for delving into the ocean of regular expressions:

- *Mastering Regular Expressions* - Understand Your Data and Be More Productive by Jeffrey E. F. Friedl, 3rd Edition, Publisher: O'Reilly Media.
- *Regular Expression Pocket Reference* by Tony Stubblebine, Publisher: O'Reilly Media.

We have shown through many examples the utility of Python in Bioinformatics studies. The applications are many and we have covered the basics of the concepts involved. The interested student is required to acquire a good mastery in Python to write and use common and specific user-defined functions for their respective scientific investigations.

REFERENCES

- *Bioinformatics Programming - Using Python* by Mitchell L. Model (O'Reilly, 2010)
- *Python Programming for Biology* by Tim J. Stevens and Wayne Boucher (Cambridge University Press, 2015).

3

SEQUENCES IN BIOLOGY

The starting point of any investigation in biology, in the domain of biomolecules and their function, is the *sequences*. **Human Genome Project** has unveiled the genome of *Homo sapiens* and the draft genomes of many living organisms (including plants, vertebrates, invertebrates, viruses, bacteria etc.) have been generated by other specialized groups and consortiums. The pace at which they are being generated has increased over the past decade. The large amount of sequence data generated in genomics experiments are housed in publicly available databases. In a parallel way, the proteome of organisms is also being studied at an ever-increasing rate. Studying the proteome and the interactions between proteins provides a dynamic angle to the sequences. We will be focusing our attention on data in such databases. We will be taking a look at some of these databases and the formats of the sequences stored in them. We will be exploring the fundamentals of sequence analysis in bioinformatics. We will interpolate some examples for getting a jump start into this ever-complex area in computation.

SEQUENCE FORMATS AND DATABASES

We will be taking up a few commonly used sequence formats. We will be dealing with any one of these in the present and subsequent chapters.

Fasta

By now any casual user in biology would have encountered FASTA files. It appears to be having an omnipresence status in the fields of computational molecular science. We had a look at it in the previous chapter in some example illustrations using Python. Briefly, the first line of a sequence entry consists of ">", followed by an identifier, which contains no whitespace. This can be followed by whitespace and a comment or description. This first line is referred to as the *comment* or *description* line. One or more *sequence data* lines may follow. The length of the sequence data lines may not be constant. Common line lengths are 60, 70, 72, and 80.

GenBank

National Center for Biotechnology Information (NCBI) maintains GenBank. It houses millions of DNA sequences. It is divided into the following sections, according to species:

a) PLN: Plant sequences

b) PRI: Primate sequences

c) ROD: Rodent sequences

d) MAM: Other mammalian sequences

e) VRT: Other vertebrate sequences

f) IVN: Invertebrate sequences

g) BCT: Bacterial sequences

h) PHG: Phage sequences

i) VRL: Other viral sequences

j) SYN: Synthetic sequences

k) UNA: Unannotated sequences

l) PAT: Patent sequences

m) NEW: New sequences

One can reach the website using the following locator:

https://www.ncbi.nlm.nih.gov/

Readers can explore the various search options. The most basic way of searching this database is by *keyword* or *sequence*. A typical entry of **GenBank** record with accession number KM399929 is shown in Figure.

LOCUS KM399929 591 bp mRNA linear PLN 01-NOV-2014

DEFINITION Arabidopsis thaliana accelerated cell death 11 mRNA, partial cds.

ACCESSION KM399929

VERSION KM399929.1

KEYWORDS .

SOURCE Arabidopsis thaliana (thale cress)

ORGANISM Arabidopsis thaliana

Eukaryota; Viridiplantae; Streptophyta; Embryophyta; Tracheophyta; Spermatophyta; Magnoliophyta; eudicotyledons; Gunneridae; Pentapetalae; rosids; malvids; Brassicales; Brassicaceae; Camelineae; Arabidopsis.

REFERENCE 1 (bases 1 to 591)

AUTHORS Zeng,L., Zhang,Q., Sun,R., Kong,H., Zhang,N. and Ma,H.

TITLE Resolution of deep angiosperm phylogeny using conserved nuclear

genes and estimates of early divergence times

JOURNAL Nat Commun 5, 4956 (2014)

PUBMED 25249442

REMARK Publication Status: Online-Only

REFERENCE 2 (bases 1 to 591)

AUTHORS Zeng,L., Zhang,Q., Sun,R., Kong,H., Zhang,N. and Ma,H.

TITLE Direct Submission

JOURNAL Submitted (26-AUG-2014) School of Life Sciences, Fudan University,

220th Handan Road, Shanghai 200433, China

COMMENT ##Assembly-Data-START##

Assembly Method :: trinity v. Trinityrnaseq_r20131110

Sequencing Technology :: Illumina

##Assembly-Data-END##

FEATURES Location/Qualifiers

source 1..591

/organism="Arabidopsis thaliana"

/mol_type="mRNA"

/db_xref="taxon:3702"

CDS <1..>591

/note="ACD11"

/codon_start=1

/product="accelerated cell death 11"

/protein_id="AIU50562.1"

/translation="KPLRKISAAFKKLAIIVNSPNPVPVTQFS HACSLVSPLFGCLGI

AFKFAEMDYVAKVDDLVRASSSISTLVVMMDKD IEADCVRKAGSHTRNLLRVKRGLDM

VKVLFEQIIASEGNSLKDPATKSYAQV FAPHHGWAIRKAVSLGMYALPTRAHLLNMLK

EDEAAAKIHMQSYVNSSAPLITYLDNLFLSKQLGIDW"

ORIGIN

1 aagccactga gaaaaatctc agccgctttc aaaaaactag caatcatcgt gaattcaccg

61 aatccggttc ctgtaacgca attctctcac gcttgctctc tggtctcgcc tctctttggt

121 tgccttggaa tagcttttaa gtttgcggaa atggactatg ttgccaaggt tgatgatctt

181 gtgagggcgt cgagttcgat atcgacatta gtggtaatga tggacaaaga tattgaggca

241 gattgtgtaa ggaaagctgg tagtcatacg agaaaccttt tgagggttaa gcgtggtctt

301 gacatggtca aagttctctt tgaacagatc atagcttccg aaggaaactc cttgaaggat

361 ccagcaacta agtcttatgc tcaagtgttt gctccccacc atggatgggc tatacggaaa

421 gctgtttctc ttgggatgta tgctcttccc acaagggctc acctacttaa tatgctcaaa

481 gaggatgagg cggcggctaa gatacatatg caaagctatg tcaattcatc ggcaccatta
541 atcacgtatc ttgataatct attcctctcc aagcaactcg gtattgattg g
//

Let us understand some of the important field definitions in the above GenBank record. We will give description of these terms in a table below.

FIELD	DESCRIPTION
LOCUS	A SHORT MNEMONIC NAME FOR THE ENTRY, CHOSEN TO SUGGEST SEQUENCE'S DEFINITION
DEFINITION	A CONCISE DESCRIPTION OF THE SEQUENCE
ACCESSION	THE PRIMARY ACCESSION NUMBER IS A UNIQUE, UNCHANGING CODE ASSIGNED TO ENTRY
VERSION	A COMPOUND IDENTIFIER CONSISTING OF THE PRIMARY ACCESSION NUMBER AND A NUMERIC VERSION NUMBER ASSOCIATED WITH THE CURRENT VERSION OF THE SEQUENCE DATA IN THE RECORD. THIS IS FOLLOWED BY AN INTEGER KEY.
KEYWORDS	SHORT PHRASES DESCRIBING GENE PRODUCTS AND OTHER INFORMATION ABOUT AN ENTRY
SOURCE	COMMON NAME OF THE ORGANISM OR THE NAME MOST FREQUENTLY USED IN THE LITERATURE
ORGANISM	FORMAL SCIENTIFIC NAME OF THE ORGANISM (FIRST LINE) AND TAXONOMIC CLASSIFICATION LEVELS
REFERENCE	CITATIONS FOR ALL ARTICLES CONTAINING DATA REPORTED IN THIS ENTRY
AUTHORS	LISTS THE AUTHORS OF THE CITATION
TITLE	FULL TITLE OF THE CITATION
JOURNAL	LISTS THE JOURNAL NAME, VOLUME, YEAR, AND PAGE NUMBERS OF THE CITATION
COMMENT	CROSS-REFERENCES TO OTHER SEQUENCE ENTRIES
FEATURES	TABLE CONTAINING INFORMATION ON PORTIONS OF THE SEQUENCE THAT CODE FOR PROTEINS AND RNA MOLECULES AND INFORMATION ON EXPERIMENTALLY DETERMINED SITES OF BIOLOGICAL SIGNIFICANCE
ORIGIN	SPECIFICATION OF HOW THE FIRST BASE OF THE REPORTED SEQUENCE IS OPERATIONALLY LOCATED WITHIN THE GENOME. WHERE POSSIBLE, THIS INCLUDES ITS LOCATION WITHIN A LARGER GENETIC MAP.S

Embl

The EMBL (**European Molecular Biology Laboratory**) is an institution maintaining several sequence repositories. It also houses DNA sequences under **Nucleotide Sequence Database**. The organization of the flat file of the sequences in the repository is very similar to that of **GenBank**. It is easy to understand the format of both EMBL and NCBI easily. A typical entry is shown in Figure below.

ID KM399929; SV 1; linear; mRNA; STD; PLN; 591 BP.

XX

AC KM399929;

XX

DT 03-NOV-2014 (Rel. 122, Created)

DT 03-NOV-2014 (Rel. 122, Last updated, Version 1)

XX

DE Arabidopsis thaliana accelerated cell death 11 mRNA, partial cds.

XX

KW .

XX

OS Arabidopsis thaliana (thale cress)

OC Eukaryota; Viridiplantae; Streptophyta; Embryophyta; Tracheophyta;

OC Spermatophyta; Magnoliophyta; eudicotyledons; Gunneridae; Pentapetalae;

OC rosids; malvids; Brassicales; Brassicaceae; Camelineae; Arabidopsis.

XX

RN [1]

RC Publication Status: Online-Only

RP 1-591

RX DOI; 10.1038/ncomms5956.

RX PUBMED; 25249442.

RA Zeng L., Zhang Q., Sun R., Kong H., Zhang N., Ma H.;

RT "Resolution of deep angiosperm phylogeny using conserved nuclear

```
genes and
RT   estimates of early divergence times";
RL   Nat Commun 5:4956-4956(2014).
XX
RN   [2]
RP   1-591
RA   Zeng L., Zhang Q., Sun R., Kong H., Zhang N., Ma H.;
RT   ;
RL   Submitted (26-AUG-2014) to the INSDC.
RL   School of Life Sciences, Fudan University, 220th Handan Road,
Shanghai
RL   200433, China
XX
DR   MD5; ab367b04b485ba27bfa4625c2727c239.
XX
CC   ##Assembly-Data-START##
CC   Assembly Method       : trinity v. Trinityrnaseq_r20131110
CC   Sequencing Technology: Illumina
CC   ##Assembly-Data-END##
XX
FH   Key             Location/Qualifiers
FH
FH   source          1..591
FT                   /organism="Arabidopsis thaliana"
FT                   /mol_type="mRNA"
FT                   /db_xref="taxon:3702"
FT   CDS             <1..>591
FT                   /codon_start=1
FT                   /product="accelerated cell death 11"
FT                   /note="ACD11"
```

FT /db_xref="GOA:O64587"

FT /db_xref="InterPro:IPR014830"

FT /db_xref="PDB:4NT1"

FT /db_xref="PDB:4NT2"

FT /db_xref="PDB:4NTG"

FT /db_xref="PDB:4NTI"

FT /db_xref="PDB:4NTO"

FT /db_xref="UniProtKB/Swiss-Prot:O64587"

FT /protein_id="AIU50562.1"

FT /translation="KPLRKISAAFKKLAIIVNSPNPVPVTQF SHACSLVSPLFGCLGIA

FT FKFAEMDYVAKVDDLVRASSSISTLVV MMDKDIEADCVRKAGSHTRNLLRVKRGLDMVK

FT VLFEQIIASEGNSLKDPATKSYAQVFAPH HGWAIRKAVSLGMYALPTRAHLLNMLKEDE

FT AAAKIHMQSYVNSSAPLITYLDNLFLSKQLGIDW"

XX

SQ Sequence 591 BP; 158 A; 126 C; 136 G; 171 T; 0 other;

aagccactga gaaaaatctc agccgctttc aaaaaactag caatcatcgt gaattcaccg 60

aatccggttc ctgtaacgca attctctcac gcttgctctc tggtctcgcc tctctttggt 120

tgccttggaa tagcttttaa gtttgcggaa atggactatg ttgccaaggt tgatgatctt 180

gtgagggcgt cgagttcgat atcgacatta gtggtaatga tggacaaaga tattgaggca 240

gattgtgtaa ggaaagctgg tagtcatacg agaaaccttt tgagggttaa gcgtggtctt 300

gacatggtca aagttctctt tgaacagatc atagcttccg aaggaaactc cttgaaggat 360

ccagcaacta agtcttatgc tcaagtgttt gctccccacc atggatgggc tatacggaaa 420

gctgtttctc ttgggatgta tgctcttccc acaagggctc acctacttaa tatgctcaaa 480

gaggatgagg cggcggctaa gatacatatg caaagctatg tcaattcatc ggcaccatta

540

atcacgtatc ttgataatct attcctctcc aagcaactcg gtattgattg g 591

//

Let us look at the field definitions.

LINE CODE	CONTENT
ID	IDENTIFICATION
AC	ACCESSION NUMBER(S)
SV	NEW SEQUENCE IDENTIFIER
DT	DATE
DE	DESCRIPTION
KW	KEYWORD
OS	ORGANISM SPECIES
OC	ORGANISM CLASSIFICATION
OG	ORGANELLE
RN	REFERENCE NUMBER
RC	REFERENCE COMMENTS
RP	REFERENCE POSITIONS
RX	REFERENCE CROSS-REFERENCES
RA	REFERENCE AUTHORS
RT	REFERENCE TITLE
RL	REFERENCE LOCATION
DR	DATABASE CROSS-REFERENCES
FH	FEATURE TABLE HEADER
FT	FEATURE TABLE DATA
CC	COMMENTS OF NOTES
XX	SPACER LINE
SQ	SEQUENCE HEADER

Pdb

The PDB (**Protein Data Bank**) is a repository of structural data of biomolecules, primarily proteins but also holds structures of DNA/RNA and complexes as well. We will be working with files taken from PDB in this book. A typical PDB file *header* looks like below (PDB Code: 4NT1).

HEADER TRANSPORT PROTEIN 29-NOV-13 4NT1

TITLE CRYSTAL STRUCTURE OF APO-FORM OF ARABIDOPSIS ACD11 (ACCELERATED-CELL-2 DEATH 11) AT 1.8 ANGSTROM RESOLUTION

COMPND MOL_ID: 1;

COMPND 2 MOLECULE: ACCELERATED-CELL-DEATH 11;

COMPND 3 CHAIN: A, B, C, D;

COMPND 4 SYNONYM: GLYCOLIPID TRANSFER PROTEIN, AT2G34690;

COMPND 5 ENGINEERED: YES

SOURCE MOL_ID: 1;

SOURCE 2 ORGANISM_SCIENTIFIC: ARABIDOPSIS THALIANA;

SOURCE 3 ORGANISM_COMMON: MOUSE-EAR CRESS;

SOURCE 4 ORGANISM_TAXID: 3702;

SOURCE 5 GENE: ACD11, AT2G34690;

SOURCE 6 EXPRESSION_SYSTEM: ESCHERICHIA COLI;

SOURCE 7 EXPRESSION_SYSTEM_TAXID: 469008;

SOURCE 8 EXPRESSION_SYSTEM_STRAIN: BL21 (DE3);

SOURCE 9 EXPRESSION_SYSTEM_VECTOR_TYPE: PLASMID;

SOURCE 10 EXPRESSION_SYSTEM_PLASMID: PET-SUMO

KEYWDS PROTEIN-LIPID COMPLEXES, GLTP-FOLD, LIPID TRANSFER PROTEIN, CELL

KEYWDS 2 DEATH, CERAMIDE-1-PHOSPHATE, C1P, LYSINE CARBOXYLATION, TRANSPORT

KEYWDS 3 PROTEIN

EXPDTA X-RAY DIFFRACTION

AUTHOR D.K.SIMANSHU, R.E.BROWN,D.J.PATEL

REVDAT 2 19-FEB-14 4NT1 1 JRNL

REVDAT 1 05-FEB-14 4NT1 0

JRNL AUTH D.K.SIMANSHU,X.ZHAI,D.MUNCH,D.HOFIUS,J.E.MARKHAM,

JRNL AUTH 2 J.BIELAWSKI,A.BIELAWSKA,L.MALININA,J.G.MOLOTKOVSKY,

JRNL AUTH 3 J.W.MUNDY, D.J.PATEL,R.E.BROWN

JRNL TITL ARABIDOPSIS ACCELERATED CELL DEATH 11, ACD11, IS A

JRNL TITL 2 CERAMIDE-1-PHOSPHATE TRANSFER PROTEIN AND INTERMEDIARY

JRNL TITL 3 REGULATOR OF PHYTOCERAMIDE LEVELS.

JRNL REF CELL REP V. 6 388 2014

JRNL REFN ESSN 2211-1247

JRNL PMID 24412362

JRNL DOI 10.1016/J.CELREP.2013.12.023

REMARK 2

REMARK 2 RESOLUTION. 1.80 ANGSTROMS.

REMARK 3

REMARK 3 REFINEMENT.

REMARK 3 PROGRAM : PHENIX (PHENIX.REFINE: 1.6.4_486)

REMARK 3 AUTHORS : PAUL ADAMS,PAVEL AFONINE,VICENT CHEN,IAN

REMARK 3 : DAVIS,KRESHNA GOPAL,RALF GROSSE-

REMARK 3 : KUNSTLEVE,LI-WEI HUNG,ROBERT IMMORMINO,

REMARK 3 : TOM IOERGER,AIRLIE MCCOY,ERIK MCKEE,NIGEL

REMARK 3 : MORIARTY,REETAL PAI,RANDY READ,JANE

REMARK 3 : RICHARDSON,DAVID RICHARDSON,TOD ROMO,JIM

```
REMARK   3            : SACCHETTINI,NICHOLAS SAUTER,JACOB
SMITH,
REMARK   3            : LAURENT STORONI,TOM TERWILLIGER,
PETER
REMARK   3            : ZWART
REMARK   3
REMARK   3   REFINEMENT TARGET: ML
REMARK   3
REMARK   3 DATA USED IN REFINEMENT.
REMARK   3   RESOLUTION RANGE HIGH (ANGSTROMS) : 1.80
REMARK   3   RESOLUTION RANGE LOW  (ANGSTROMS) : 33.82
REMARK   3   MIN(FOBS/SIGMA_FOBS)              : 0.000
REMARK   3   COMPLETENESS FOR RANGE        (%) : 95.0
REMARK   3   NUMBER OF REFLECTIONS             : 75108
REMARK   3
REMARK   3  FIT TO DATA USED IN REFINEMENT.
REMARK   3   R VALUE     (WORKING + TEST SET) : 0.176
REMARK   3   R VALUE            (WORKING SET) : 0.174
REMARK   3   FREE R VALUE                     : 0.213
REMARK   3   FREE R VALUE TEST SET SIZE   (%) : 5.030
REMARK   3   FREE R VALUE TEST SET COUNT      : 3780
REMARK   3
REMARK   3  FIT TO DATA USED IN REFINEMENT (IN BINS).
REMARK   3    BIN RESOLUTION RANGE COMPL.    NWORK
NFREE   RWORK RFREE
REMARK   3     1 33.8286 - 3.8762    0.98     7696   433  0.1619 0.1873
REMARK   3     2  3.8762 -  3.0773   0.99     7526   380  0.1681 0.1975
REMARK   3     3  3.0773 -  2.6885   0.99     7445   429  0.1894 0.2266
REMARK   3     4  2.6885 -  2.4428   0.98     7390   364  0.1875 0.2351
REMARK   3     5 2.4428 -  2.2677    0.98     7290   391  0.1800 0.2292
REMARK   3     6 2.2677 -  2.1341    0.97     7258   371  0.1685 0.2137
```

REMARK 3 7 2.1341 - 2.0272 0.97 7153 400 0.1703 0.2359

REMARK 3 8 2.0272 - 1.9390 0.94 7029 368 0.1832 0.2422

REMARK 3 9 1.9390 - 1.8643 0.90 6659 354 0.1798 0.2440

REMARK 3 10 1.8643 - 1.8000 0.79 5882 290 0.1966 0.2606

REMARK 3

REMARK 3 BULK SOLVENT MODELLING.

REMARK 3 METHOD USED : FLAT BULK SOLVENT MODEL

REMARK 3 SOLVENT RADIUS : 1.20

REMARK 3 SHRINKAGE RADIUS : 0.95

REMARK 3 K_SOL : 0.35

REMARK 3 B_SOL : 44.89

REMARK 3

REMARK 3 ERROR ESTIMATES.

REMARK 3 COORDINATE ERROR (MAXIMUM-LIKELIHOOD BASED) : 0.200

REMARK 3 PHASE ERROR (DEGREES, MAXIMUM-LIKELIHOOD BASED) : 21.170

REMARK 3

REMARK 3 B VALUES.

REMARK 3 FROM WILSON PLOT (A**2) : NULL

REMARK 3 MEAN B VALUE (OVERALL, A**2) : 27.27

REMARK 3 OVERALL ANISOTROPIC B VALUE.

REMARK 3 B11 (A**2) : 5.20180

REMARK 3 B22 (A**2) : -2.89340

REMARK 3 B33 (A**2) : -2.30840

REMARK 3 B12 (A**2) : -0.00000

REMARK 3 B13 (A**2) : -0.00000

REMARK 3 B23 (A**2) : -0.00000

REMARK 3

REMARK 3 TWINNING INFORMATION.

```
REMARK   3   FRACTION: NULL
REMARK   3   OPERATOR: NULL
REMARK   3
REMARK   3  DEVIATIONS FROM IDEAL VALUES.
REMARK   3                RMSD       COUNT
REMARK   3   BOND      : 0.006        6363
REMARK   3   ANGLE     : 0.968        8641
REMARK   3   CHIRALITY :  0.066        1013
REMARK   3   PLANARITY :  0.005        1091
REMARK   3   DIHEDRAL  : 11.592        2383
REMARK   3
REMARK   3  TLS DETAILS
REMARK   3   NUMBER OF TLS GROUPS  : NULL
REMARK   3
REMARK   3  NCS DETAILS
REMARK   3   NUMBER OF NCS GROUPS : NULL
REMARK   3
REMARK   3  OTHER REFINEMENT REMARKS: NULL
REMARK   4
REMARK   4 4NT1 COMPLIES WITH FORMAT V. 3.30, 13-JUL-11
REMARK 100
REMARK 100 THIS ENTRY HAS BEEN PROCESSED BY RCSB ON
16-DEC-13.
REMARK 100 THE RCSB ID CODE IS RCSB083619.
REMARK 200
REMARK 200 EXPERIMENTAL DETAILS
REMARK 200  EXPERIMENT TYPE              : X-RAY DIFFRACTION
REMARK 200  DATE OF DATA COLLECTION       : 04-FEB-08
REMARK 200  TEMPERATURE          (KELVIN) : 100
REMARK 200  PH                           : 8.5
```

REMARK 200 NUMBER OF CRYSTALS USED : 1

REMARK 200

REMARK 200 SYNCHROTRON (Y/N) : Y

REMARK 200 RADIATION SOURCE : NSLS

REMARK 200 BEAMLINE : X29A

REMARK 200 X-RAY GENERATOR MODEL : NULL

REMARK 200 MONOCHROMATIC OR LAUE (M/L) : M

REMARK 200 WAVELENGTH OR RANGE (A) : 0.990

REMARK 200 MONOCHROMATOR : DOUBLE CRYSTAL SI(111)

REMARK 200 OPTICS : NULL

REMARK 200

REMARK 200 DETECTOR TYPE : CCD

REMARK 200 DETECTOR MANUFACTURER : ADSC QUANTUM 315

REMARK 200 INTENSITY-INTEGRATION SOFTWARE : HKL-2000

REMARK 200 DATA SCALING SOFTWARE : SCALEPACK

REMARK 200

REMARK 200 NUMBER OF UNIQUE REFLECTIONS : 79607

REMARK 200 RESOLUTION RANGE HIGH (A) : 1.800

REMARK 200 RESOLUTION RANGE LOW (A) : 50.000

REMARK 200 REJECTION CRITERIA (SIGMA(I)) : NULL

REMARK 200

REMARK 200 OVERALL.

REMARK 200 COMPLETENESS FOR RANGE (%) : 97.8

REMARK 200 DATA REDUNDANCY : 4.500

REMARK 200 R MERGE (I) : 0.08000

REMARK 200 R SYM (I) : NULL

REMARK 200 <I/SIGMA(I)> FOR THE DATA SET : 26.2000

REMARK 200

REMARK 200 IN THE HIGHEST RESOLUTION SHELL.

REMARK 200 HIGHEST RESOLUTION SHELL, RANGE HIGH (A) : 1.80

REMARK 200 HIGHEST RESOLUTION SHELL, RANGE LOW (A) : 1.86

REMARK 200 COMPLETENESS FOR SHELL (%) : NULL

REMARK 200 DATA REDUNDANCY IN SHELL : 3.60

REMARK 200 R MERGE FOR SHELL (I) : 0.30900

REMARK 200 R SYM FOR SHELL (I) : NULL

REMARK 200 <I/SIGMA(I)> FOR SHELL : 2.900

REMARK 200

REMARK 200 DIFFRACTION PROTOCOL: SINGLE WAVELENGTH

REMARK 200 METHOD USED TO DETERMINE THE STRUCTURE: SIRAS

REMARK 200 SOFTWARE USED: SHARP

REMARK 200 STARTING MODEL: NULL

REMARK 200

REMARK 200 REMARK: NULL

REMARK 280

REMARK 280 CRYSTAL

REMARK 280 SOLVENT CONTENT, VS (%): 46.68

REMARK 280 MATTHEWS COEFFICIENT, VM (ANGSTROMS**3/DA): 2.31

REMARK 280

REMARK 280 CRYSTALLIZATION CONDITIONS: 0.1 M TRIS, PH 8.5, 1.2 M POTASSIUM

REMARK 280 SODIUM TARTRATE TETRAHYDRATE, VAPOR DIFFUSION, HANGING DROP,

REMARK 280 TEMPERATURE 293K

REMARK 290

REMARK 290 CRYSTALLOGRAPHIC SYMMETRY

REMARK 290 SYMMETRY OPERATORS FOR SPACE GROUP: P 21 21 21

REMARK 290

REMARK 290 SYMOP SYMMETRY

REMARK 290 NNNMMM OPERATOR

REMARK 290 1555 X,Y,Z

REMARK 290 2555 -X+1/2,-Y,Z+1/2

REMARK 290 3555 -X,Y+1/2,-Z+1/2

REMARK 290 4555 X+1/2,-Y+1/2,-Z

REMARK 290

REMARK 290 WHERE NNN -> OPERATOR NUMBER

REMARK 290 MMM -> TRANSLATION VECTOR

REMARK 290

REMARK 290 CRYSTALLOGRAPHIC SYMMETRY TRANSFORMATIONS

REMARK 290 THE FOLLOWING TRANSFORMATIONS OPERATE ON THE ATOM/HETATM

REMARK 290 RECORDS IN THIS ENTRY TO PRODUCE CRYSTALLOGRAPHICALLY

REMARK 290 RELATED MOLECULES.

REMARK 290 SMTRY1 1 1.000000 0.000000 0.000000 0.00000

REMARK 290 SMTRY2 1 0.000000 1.000000 0.000000 0.00000

REMARK 290 SMTRY3 1 0.000000 0.000000 1.000000 0.00000

REMARK 290 SMTRY1 2 -1.000000 0.000000 0.000000 35.37800

REMARK 290 SMTRY2 2 0.000000 -1.000000 0.000000 0.00000

REMARK 290 SMTRY3 2 0.000000 0.000000 1.000000 77.35800

REMARK 290 SMTRY1 3 -1.000000 0.000000 0.000000 0.00000

REMARK 290 SMTRY2 3 0.000000 1.000000 0.000000 38.50700

REMARK 290 SMTRY3 3 0.000000 0.000000 -1.000000 77.35800

REMARK 290 SMTRY1 4 1.000000 0.000000 0.000000 35.37800

REMARK 290 SMTRY2 4 0.000000 -1.000000 0.000000 38.50700

REMARK 290 SMTRY3 4 0.000000 0.000000 -1.000000 0.00000

REMARK 290

REMARK 290 REMARK: NULL

REMARK 300

REMARK 300 BIOMOLECULE: 1, 2, 3, 4

REMARK 300 SEE REMARK 350 FOR THE AUTHOR PROVIDED AND/OR PROGRAM

REMARK 300 GENERATED ASSEMBLY INFORMATION FOR THE STRUCTURE IN

REMARK 300 THIS ENTRY. THE REMARK MAY ALSO PROVIDE INFORMATION ON

REMARK 300 BURIED SURFACE AREA.

REMARK 350

REMARK 350 COORDINATES FOR A COMPLETE MULTIMER REPRESENTING THE KNOWN

REMARK 350 BIOLOGICALLY SIGNIFICANT OLIGOMERIZATION STATE OF THE

REMARK 350 MOLECULE CAN BE GENERATED BY APPLYING BIOMT TRANSFORMATIONS

REMARK 350 GIVEN BELOW. BOTH NON-CRYSTALLOGRAPHIC AND

REMARK 350 CRYSTALLOGRAPHIC OPERATIONS ARE GIVEN.

REMARK 350

REMARK 350 BIOMOLECULE: 1

REMARK 350 AUTHOR DETERMINED BIOLOGICAL UNIT: MONOMERIC

REMARK 350 SOFTWARE DETERMINED QUATERNARY STRUCTURE: MONOMERIC

REMARK 350 SOFTWARE USED: PISA

REMARK 350 APPLY THE FOLLOWING TO CHAINS: A

REMARK 350 BIOMT1 1 1.000000 0.000000 0.000000 0.00000

REMARK 350 BIOMT2 1 0.000000 1.000000 0.000000 0.00000

REMARK 350 BIOMT3 1 0.000000 0.000000 1.000000 0.00000

REMARK 350

REMARK 350 BIOMOLECULE: 2

REMARK 350 AUTHOR DETERMINED BIOLOGICAL UNIT: MONOMERIC

REMARK 350 SOFTWARE DETERMINED QUATERNARY STRUCTURE: MONOMERIC

REMARK 350 SOFTWARE USED: PISA

REMARK 350 APPLY THE FOLLOWING TO CHAINS: B

REMARK 350 BIOMT1 1 1.000000 0.000000 0.000000 0.00000

REMARK 350 BIOMT2 1 0.000000 1.000000 0.000000 0.00000

REMARK 350 BIOMT3 1 0.000000 0.000000 1.000000 0.00000

REMARK 350

REMARK 350 BIOMOLECULE: 3

REMARK 350 AUTHOR DETERMINED BIOLOGICAL UNIT: MONOMERIC

REMARK 350 SOFTWARE DETERMINED QUATERNARY STRUCTURE: MONOMERIC

REMARK 350 SOFTWARE USED: PISA

REMARK 350 APPLY THE FOLLOWING TO CHAINS: C

REMARK 350 BIOMT1 1 1.000000 0.000000 0.000000 0.00000

REMARK 350 BIOMT2 1 0.000000 1.000000 0.000000 0.00000

REMARK 350 BIOMT3 1 0.000000 0.000000 1.000000 0.00000

REMARK 350

REMARK 350 BIOMOLECULE: 4

REMARK 350 AUTHOR DETERMINED BIOLOGICAL UNIT: MONOMERIC

REMARK 350 SOFTWARE DETERMINED QUATERNARY STRUCTURE: MONOMERIC

REMARK 350 SOFTWARE USED: PISA

REMARK 350 APPLY THE FOLLOWING TO CHAINS: D

REMARK 350 BIOMT1 1 1.000000 0.000000 0.000000 0.00000

REMARK 350 BIOMT2 1 0.000000 1.000000 0.000000 0.00000

REMARK 350 BIOMT3 1 0.000000 0.000000 1.000000 0.00000

REMARK 465

REMARK 465 MISSING RESIDUES

REMARK 465 THE FOLLOWING RESIDUES WERE NOT LOCATED IN THE

REMARK 465 EXPERIMENT. (M=MODEL NUMBER; RES=RESIDUE NAME; C=CHAIN

REMARK 465 IDENTIFIER; SSSEQ=SEQUENCE NUMBER; I=INSERTION CODE.)

REMARK 465

REMARK 465 M RES C SSSEQI

REMARK 465 SER A 0

REMARK 465 MET A 1

REMARK 465 ALA A 2

REMARK 465 ASP A 3

REMARK 465 SER A 4

REMARK 465 GLU A 5

REMARK 465 ALA A 6

REMARK 465 SER B 0

REMARK 465 MET B 1

REMARK 465 ALA B 2

REMARK 465 ASP B 3

REMARK 465 SER B 4

REMARK 465 GLU B 5

REMARK 465 ALA B 6

REMARK 465 SER C 0

REMARK 465 MET C 1

REMARK 465 ALA C 2

REMARK 465 ASP C 3

REMARK 465 SER C 4

REMARK 465 GLU C 5

REMARK 465 ALA C 6

REMARK 465 SER D 0

REMARK 465 MET D 1

REMARK 465 ALA D 2

REMARK 465 ASP D 3

REMARK 465 SER D 4

REMARK 465 GLU D 5

REMARK 465 GLN D 201

REMARK 465 LEU D 202

REMARK 465 GLY D 203

REMARK 465 ILE D 204

REMARK 465 ASP D 205

REMARK 465 TRP D 206

REMARK 470

REMARK 470 MISSING ATOM

REMARK 470 THE FOLLOWING RESIDUES HAVE MISSING ATOMS (M=MODEL NUMBER;

REMARK 470 RES=RESIDUE NAME; C=CHAIN IDENTIFIER; SSEQ=SEQUENCE NUMBER;

REMARK 470 I=INSERTION CODE):

REMARK 470 M RES CSSEQI ATOMS

REMARK 470 ASP A 7 CG OD1 OD2

REMARK 470 ASN A 28 CG OD1 ND2

REMARK 470 KCX A 64 CX OQ1 OQ2

REMARK 470 TYR A 156 CG CD1 CD2 CE1 CE2 CZ OH

REMARK 470 ASP B 7 CG OD1 OD2

REMARK 470 ASP C 7 CG OD1 OD2

REMARK 470 LYS C 128 CD CE NZ

REMARK 470 ASP D 7 CG OD1 OD2

REMARK 470 KCX D 64 CX OQ1 OQ2

REMARK 470 ARG D 92 CG CD NE CZ NH1 NH2

REMARK 500

REMARK 500 GEOMETRY AND STEREOCHEMISTRY

REMARK 500 SUBTOPIC: TORSION ANGLES

REMARK 500

REMARK 500 TORSION ANGLES OUTSIDE THE EXPECTED RAMACHANDRAN REGIONS:

REMARK 500 (M=MODEL NUMBER; RES=RESIDUE NAME; C=CHAIN IDENTIFIER;

REMARK 500 SSEQ=SEQUENCE NUMBER; I=INSERTION CODE).

REMARK 500

REMARK 500 STANDARD TABLE:

REMARK 500 FORMAT:(10X,I3,1X,A3,1X,A1,I4,A1,4X,F7.2,3X ,F7.2)

REMARK 500

REMARK 500 EXPECTED VALUES: GJ KLEYWEGT AND TA JONES (1996). PHI/PSI-

REMARK 500 CHOLOGY: RAMACHANDRAN REVISITED. STRUCTURE 4, 1395 - 1400

REMARK 500

REMARK 500 M RES CSSEQI PSI PHI

REMARK 500 TYR A 61 -59.84 -127.05

REMARK 500 LYS B 8 121.70 -38.88

REMARK 500 TYR B 61 -63.81 -126.22

REMARK 500 SER B 76 -75.65 -69.77

REMARK 500 TYR C 61 -68.21 -127.31

REMARK 500 TYR D 61 -62.52 -123.46

REMARK 500 SER D 199 -6.43 -58.30

REMARK 500

REMARK 500 REMARK: NULL

REMARK 525

REMARK 525 SOLVENT

REMARK 525

REMARK 525 THE SOLVENT MOLECULES HAVE CHAIN IDENTIFIERS THAT

REMARK 525 INDICATE THE POLYMER CHAIN WITH WHICH THEY ARE MOST

REMARK 525 CLOSELY ASSOCIATED. THE REMARK LISTS ALL THE SOLVENT

REMARK 525 MOLECULES WHICH ARE MORE THAN 5A AWAY FROM THE

REMARK 525 NEAREST POLYMER CHAIN (M = MODEL NUMBER;

REMARK 525 RES=RESIDUE NAME; C=CHAIN IDENTIFIER; SSEQ=SEQUENCE

REMARK 525 NUMBER; I=INSERTION CODE):

REMARK 525

REMARK 525 M RES CSSEQI

REMARK 525 HOH B 462 DISTANCE = 5.03 ANGSTROMS

REMARK 620

REMARK 620 METAL COORDINATION

REMARK 620 (M=MODEL NUMBER; RES=RESIDUE NAME; C=CHAIN IDENTIFIER;

REMARK 620 SSEQ=SEQUENCE NUMBER; I=INSERTION CODE):

REMARK 620

REMARK 620 COORDINATION ANGLES FOR: M RES CSSEQI METAL

REMARK 620 NA A 301 NA

REMARK 620 N RES CSSEQI ATOM

REMARK 620 1 SER A 72 O

REMARK 620 2 THR A 34 OG1 81.1

REMARK 620 3 ILE A 75 O 76.3 125.7

REMARK 620 4 HOH A 401 O 122.0 77.7 74.0

REMARK 620 5 HOH A 508 O 142.0 136.9 79.0 77.0

REMARK 620 N 1 2 3 4

REMARK 800

REMARK 800 SITE

REMARK 800 SITE_IDENTIFIER: AC1

REMARK 800 EVIDENCE_CODE: SOFTWARE

REMARK 800 SITE_DESCRIPTION: BINDING SITE FOR RESIDUE NA A 301

REMARK 900

REMARK 900 RELATED ENTRIES

REMARK 900 RELATED ID: 4NT2 RELATED DB: PDB

REMARK 900 RELATED ID: 4NTG RELATED DB: PDB

REMARK 900 RELATED ID: 4NTI RELATED DB: PDB

REMARK 900 RELATED ID: 4NTO RELATED DB: PDB

DBREF 4NT1 A 1 206 UNP O64587 O64587_ARATH 1 206

DBREF 4NT1 B 1 206 UNP O64587 O64587_ARATH 1 206

DBREF 4NT1 C 1 206 UNP O64587 O64587_ARATH 1 206

DBREF 4NT1 D 1 206 UNP O64587 O64587_ARATH 1 206

SEQADV 4NT1 SER A 0 UNP O64587 EXPRESSION TAG

SEQADV 4NT1 SER B 0 UNP O64587 EXPRESSION TAG

SEQADV 4NT1 SER C 0 UNP O64587 EXPRESSION TAG

SEQADV 4NT1 SER D 0 UNP O64587 EXPRESSION TAG

SEQRES 1 A 207 SER MET ALA ASP SER GLU ALA ASP LYS PRO LEU ARG LYS

SEQRES 2 A 207 ILE SER ALA ALA PHE LYS LYS LEU ALA ILE ILE VAL ASN

SEQRES 3 A 207 SER PRO ASN PRO GLU VAL PRO VAL THR GLN PHE SER HIS

SEQRES 4 A 207 ALA CYS SER LEU VAL SER PRO LEU PHE GLY CYS LEU GLY

SEQRES 5 A 207 ILE ALA PHE LYS PHE ALA GLU MET ASP TYR VAL ALA KCX

SEQRES 6 A 207 VAL ASP ASP LEU VAL ARG ALA SER SER SER ILE SER THR

SEQRES 7 A 207 LEU VAL VAL MET MET ASP LYS ASP ILE GLU ALA ASP CYS

SEQRES 8 A 207 VAL ARG LYS ALA GLY SER HIS THR ARG ASN LEU LEU ARG

SEQRES 9 A 207 VAL LYS ARG GLY LEU ASP MET VAL LYS VAL LEU PHE GLU

SEQRES 10 A 207 GLN ILE ILE ALA SER GLU GLY ASP ASN SER LEU LYS ASP

SEQRES 11 A 207 PRO ALA THR LYS SER TYR ALA GLN VAL PHE ALA PRO HIS

SEQRES 12 A 207 HIS GLY TRP ALA ILE ARG LYS ALA VAL SER LEU GLY MET

SEQRES 13 A 207 TYR ALA LEU PRO THR ARG ALA HIS LEU LEU ASN MET LEU

SEQRES 14 A 207 LYS GLU ASP GLU ALA ALA ALA LYS ILE HIS MET GLN SER

SEQRES 15 A 207 TYR VAL ASN SER SER ALA PRO LEU ILE THR TYR LEU ASP

SEQRES 16 A 207 ASN LEU PHE LEU SER LYS GLN LEU GLY ILE ASP TRP

SEQRES 1 B 207 SER MET ALA ASP SER GLU ALA ASP LYS PRO LEU ARG LYS

SEQRES 2 B 207 ILE SER ALA ALA PHE LYS LYS LEU ALA ILE ILE VAL ASN

SEQRES 3 B 207 SER PRO ASN PRO GLU VAL PRO VAL THR GLN PHE SER HIS

SEQRES 4 B 207 ALA CYS SER LEU VAL SER PRO LEU PHE GLY CYS LEU GLY

SEQRES 5 B 207 ILE ALA PHE LYS PHE ALA GLU MET ASP TYR VAL ALA KCX

SEQRES 6 B 207 VAL ASP ASP LEU VAL ARG ALA SER SER SER ILE SER THR

SEQRES 7 B 207 LEU VAL VAL MET MET ASP LYS ASP ILE GLU ALA ASP CYS

SEQRES 8 B 207 VAL ARG LYS ALA GLY SER HIS THR ARG ASN LEU LEU ARG

SEQRES 9 B 207 VAL LYS ARG GLY LEU ASP MET VAL LYS VAL LEU PHE GLU

SEQRES 10 B 207 GLN ILE ILE ALA SER GLU GLY ASP ASN SER LEU LYS ASP

SEQRES 11 B 207 PRO ALA THR LYS SER TYR ALA GLN VAL PHE ALA PRO HIS

SEQRES 12 B 207 HIS GLY TRP ALA ILE ARG LYS ALA VAL SER LEU GLY MET

SEQRES 13 B 207 TYR ALA LEU PRO THR ARG ALA HIS LEU LEU ASN MET LEU

SEQRES 14 B 207 LYS GLU ASP GLU ALA ALA ALA LYS ILE HIS MET GLN SER

SEQRES 15 B 207 TYR VAL ASN SER SER ALA PRO LEU ILE THR TYR LEU ASP

SEQRES 16 B 207 ASN LEU PHE LEU SER LYS GLN LEU GLY ILE ASP TRP

SEQRES 1 C 207 SER MET ALA ASP SER GLU ALA ASP LYS PRO LEU ARG LYS

SEQRES 2 C 207 ILE SER ALA ALA PHE LYS LYS LEU ALA ILE ILE VAL ASN

SEQRES 3 C 207 SER PRO ASN PRO GLU VAL PRO VAL THR GLN PHE SER HIS

SEQRES 4 C 207 ALA CYS SER LEU VAL SER PRO LEU PHE GLY CYS LEU GLY

SEQRES 5 C 207 ILE ALA PHE LYS PHE ALA GLU MET ASP TYR VAL ALA KCX

SEQRES 6 C 207 VAL ASP ASP LEU VAL ARG ALA SER SER SER ILE SER THR

SEQRES 7 C 207 LEU VAL VAL MET MET ASP LYS ASP ILE GLU ALA ASP CYS

SEQRES 8 C 207 VAL ARG LYS ALA GLY SER HIS THR ARG ASN LEU LEU ARG

SEQRES 9 C 207 VAL LYS ARG GLY LEU ASP MET VAL LYS VAL LEU PHE GLU

SEQRES 10 C 207 GLN ILE ILE ALA SER GLU GLY ASP ASN SER LEU LYS ASP

SEQRES 11 C 207 PRO ALA THR LYS SER TYR ALA GLN VAL PHE ALA PRO HIS

SEQRES 12 C 207 HIS GLY TRP ALA ILE ARG LYS ALA VAL SER LEU GLY MET

SEQRES 13 C 207 TYR ALA LEU PRO THR ARG ALA HIS LEU LEU ASN MET LEU

SEQRES 14 C 207 LYS GLU ASP GLU ALA ALA ALA LYS ILE HIS MET GLN SER

SEQRES 15 C 207 TYR VAL ASN SER SER ALA PRO LEU ILE THR TYR LEU ASP

SEQRES 16 C 207 ASN LEU PHE LEU SER LYS GLN LEU GLY ILE ASP TRP

SEQRES 1 D 207 SER MET ALA ASP SER GLU ALA ASP LYS PRO LEU ARG LYS

SEQRES 2 D 207 ILE SER ALA ALA PHE LYS LYS LEU ALA ILE ILE VAL ASN

SEQRES 3 D 207 SER PRO ASN PRO GLU VAL PRO VAL THR GLN PHE SER HIS

SEQRES 4 D 207 ALA CYS SER LEU VAL SER PRO LEU PHE GLY CYS LEU GLY

SEQRES 5 D 207 ILE ALA PHE LYS PHE ALA GLU MET ASP TYR VAL ALA KCX

SEQRES 6 D 207 VAL ASP ASP LEU VAL ARG ALA SER SER SER ILE SER THR

SEQRES 7 D 207 LEU VAL VAL MET MET ASP LYS ASP ILE GLU ALA ASP CYS

SEQRES 8 D 207 VAL ARG LYS ALA GLY SER HIS THR ARG ASN LEU LEU ARG

SEQRES 9 D 207 VAL LYS ARG GLY LEU ASP MET VAL LYS VAL LEU PHE GLU

SEQRES 10 D 207 GLN ILE ILE ALA SER GLU GLY ASP ASN SER LEU LYS ASP

SEQRES 11 D 207 PRO ALA THR LYS SER TYR ALA GLN VAL PHE ALA PRO HIS

SEQRES 12 D 207 HIS GLY TRP ALA ILE ARG LYS ALA VAL SER LEU GLY MET

SEQRES 13 D 207 TYR ALA LEU PRO THR ARG ALA HIS LEU LEU ASN MET LEU

SEQRES 14 D 207 LYS GLU ASP GLU ALA ALA ALA LYS ILE HIS MET GLN SER

SEQRES 15 D 207 TYR VAL ASN SER SER ALA PRO LEU ILE THR TYR LEU ASP

SEQRES 16 D 207 ASN LEU PHE LEU SER LYS GLN LEU GLY ILE ASP TRP

MODRES 4NT1 KCX A 64 LYS LYSINE NZ-CARBOXYLIC ACID

MODRES 4NT1 KCX B 64 LYS LYSINE NZ-CARBOXYLIC ACID

MODRES 4NT1 KCX C 64 LYS LYSINE NZ-CARBOXYLIC ACID

MODRES 4NT1 KCX D 64 LYS LYSINE NZ-CARBOXYLIC ACID

HET KCX A 64 9

HET KCX B 64 12

HET KCX C 64 12

HET KCX D 64 9

HET NA A 301 1

HETNAM KCX LYSINE NZ-CARBOXYLIC ACID

HETNAM NA SODIUM ION

FORMUL 1 KCX 4(C7 H14 N2 O4)

```
FORMUL   5  NA    NA 1+
FORMUL   6  HOH   *753(H2 O)
HELIX    1   1 LYS A    8  ASN A   25  1                18
HELIX    2   2 PRO A   32  GLY A   51  1                20
HELIX    3   3 ILE A   52  LYS A   55  5                4
HELIX    4   4 PHE A   56  TYR A   61  1                6
HELIX    5   5 TYR A   61  SER A   72  1                12
HELIX    6   6 SER A   73  ILE A   75  5                3
HELIX    7   7 THR A   77  ALA A   88  1                12
HELIX    8   8 SER A   96  SER A  121  1                26
HELIX    9   9 LEU A  127  PHE A  139  1                13
HELIX   10  10 ALA A  140  HIS A  143  5                4
HELIX   11  11 GLY A  144  MET A  155  1                12
HELIX   12  12 TYR A  156  LEU A  158  5                3
HELIX   13  13 THR A  160  LEU A  168  1                9
HELIX   14  14 ASP A  171  SER A  199  1                29
HELIX   15  15 LYS B    8  ASN B   25  1                18
HELIX   16  16 PRO B   32  LEU B   50  1                19
HELIX   17  17 GLY B   51  PHE B   54  5                4
HELIX   18  18 LYS B   55  TYR B   61  1                7
HELIX   19  19 TYR B   61  SER B   73  1                13
HELIX   20  20 THR B   77  ASP B   89  1                13
HELIX   21  21 SER B   96  SER B  121  1                26
HELIX   22  22 LEU B  127  PHE B  139  1                13
HELIX   23  23 ALA B  140  HIS B  143  5                4
HELIX   24  24 GLY B  144  GLY B  154  1                11
HELIX   25  25 MET B  155  LEU B  158  5                4
HELIX   26  26 THR B  160  LEU B  168  1                9
HELIX   27  27 ASP B  171  LYS B  200  1                30
HELIX   28  28 LYS C    8  ASN C   25  1                18
```

```
HELIX   29  29 PRO C   32  GLY C   51  1                                  20
HELIX   30  30 ILE C   52  LYS C   55  5                                   4
HELIX   31  31 PHE C   56  TYR C   61  1                                   6
HELIX   32  32 TYR C   61  SER C   72  1                                  12
HELIX   33  33 THR C   77  ASP C   89  1                                  13
HELIX   34  34 SER C   96  SER C  121  1                                  26
HELIX   35  35 LEU C  127  PHE C  139  1                                  13
HELIX   36  36 ALA C  140  HIS C  143  5                                   4
HELIX   37  37 GLY C  144  MET C  155  1                                  12
HELIX   38  38 TYR C  156  LEU C  158  5                                   3
HELIX   39  39 THR C  160  LYS C  169  1                                  10
HELIX   40  40 ASP C  171  LYS C  200  1                                  30
HELIX   41  41 LYS D    8  ASN D   25  1                                  18
HELIX   42  42 PRO D   32  SER D   44  1                                  13
HELIX   43  43 SER D   44  LEU D   50  1                                   7
HELIX   44  44 GLY D   51  LYS D   55  5                                   5
HELIX   45  45 PHE D   56  TYR D   61  1                                   6
HELIX   46  46 TYR D   61  SER D   72  1                                  12
HELIX   47  47 THR D   77  ALA D   88  1                                  12
HELIX   48  48 SER D   96  ALA D  120  1                                  25
HELIX   49  49 LEU D  127  PHE D  139  1                                  13
HELIX   50  50 ALA D  140  HIS D  143  5                                   4
HELIX   51  51 GLY D  144  MET D  155  1                                  12
HELIX   52  52 TYR D  156  LEU D  158  5                                   3
HELIX   53  53 THR D  160  LYS D  169  1                                  10
HELIX   54  54 ASP D  171  SER D  199  1                                  29
LINK         C   ALA A  63                 N   KCX A  64     1555   1555  1.33
LINK         C   KCX A  64                 N   VAL A  65     1555   1555  1.33
LINK         C   ALA B  63                 N   KCX B  64     1555   1555  1.33
LINK         C   KCX B  64                 N   VAL B  65     1555   1555  1.33
```

```
LINK     C   ALA C  63      N   KCX C  64    1555  1555  1.33
LINK     C   KCX C  64      N   VAL C  65    1555  1555  1.33
LINK     C   ALA D  63      N   KCX D  64    1555  1555  1.33
LINK     C   KCX D  64      N   VAL D  65    1555  1555  1.33
LINK     O   SER A  72      NA   NA A 301    1555  1555  2.48
LINK     OG1 THR A  34      NA   NA A 301    1555  1555  2.62
LINK     O   ILE A  75      NA   NA A 301    1555  1555  2.79
LINK     NA   NA A 301      O   HOH A 401    1555  1555  2.32
LINK     NA   NA A 301      O   HOH A 508    1555  1555  2.76
SITE   1 AC1 5 THR A  34  SER A  72  ILE A  75  HOH A 401
SITE   2 AC1 5 HOH A 508
CRYST1  70.756  77.014 154.716 90.00 90.00 90.00 P 21 21 21  16
ORIGX1    1.000000  0.000000  0.000000     0.00000
ORIGX2    0.000000  1.000000  0.000000     0.00000
ORIGX3    0.000000  0.000000  1.000000     0.00000
SCALE1    0.014133  0.000000  0.000000     0.00000
SCALE2    0.000000  0.012985  0.000000     0.00000
SCALE3    0.000000  0.000000  0.006463     0.00000
```

What the different fields mean in all the above sequence files are described in the respective websites. The rest of the file will contain the coordinates of each atom as observed by solving the crystal structure. A sample is given below for the first four residues in the crystal structure of *apo*-form of *Arabidopsis* ACD11 (accelerated-cell-2 death 11).

```
ATOM   1  N   ASP A  7   18.457  48.649  27.083  1.00 49.40   N
ATOM   2  CA  ASP A  7   18.910  47.354  26.580  1.00 51.36   C
ATOM   3  C   ASP A  7   19.344  46.436  27.719  1.00 52.33   C
ATOM   4  O   ASP A  7   18.637  45.488  28.068  1.00 51.62   O
ATOM   5  CB  ASP A  7   20.060  47.528  25.584  1.00 57.80   C
ATOM   6  N   LYS A  8   20.511  46.717  28.293  1.00 41.23   N
ATOM   7  CA  LYS A  8   21.039  45.904  29.387  1.00 38.94   C
ATOM   8  C   LYS A  8   21.417  46.775  30.588  1.00 37.69   C
```

ATOM 9 O LYS A 8 22.576 47.165 30.740 1.00 30.35 O

ATOM 10 CB LYS A 8 22.254 45.102 28.919 1.00 38.85 C

ATOM 11 CG LYS A 8 21.977 44.125 27.782 1.00 40.46 C

ATOM 12 CD LYS A 8 20.998 43.037 28.202 1.00 36.15 C

ATOM 13 CE LYS A 8 20.939 41.924 27.165 1.00 39.84 C

ATOM 14 NZ LYS A 8 19.822 40.969 27.416 1.00 43.07 N

ATOM 15 N PRO A 9 20.436 47.079 31.447 1.00 31.03 N

ATOM 16 CA PRO A 9 20.649 48.020 32.557 1.00 31.00 C

ATOM 17 C PRO A 9 21.762 47.614 33.525 1.00 25.87 C

ATOM 18 O PRO A 9 22.487 48.490 34.000 1.00 25.06 O

ATOM 19 CB PRO A 9 19.288 48.049 33.271 1.00 35.00 C

ATOM 20 CG PRO A 9 18.539 46.846 32.755 1.00 36.53 C

ATOM 21 CD PRO A 9 19.039 46.622 31.365 1.00 36.64 C

ATOM 22 N LEU A 10 21.897 46.323 33.829 1.00 23.87 N

ATOM 23 CA LEU A 10 22.977 45.881 34.709 1.00 26.91 C

ATOM 24 C LEU A 10 24.335 46.113 34.067 1.00 23.85 C

ATOM 25 O LEU A 10 25.327 46.315 34.756 1.00 21.87 O

ATOM 26 CB LEU A 10 22.854 44.394 35.042 1.00 22.72 C

ATOM 27 CG LEU A 10 21.987 44.020 36.240 1.00 31.89 C

ATOM 28 CD1 LEU A 10 22.114 42.525 36.531 1.00 29.03 C

ATOM 29 CD2 LEU A 10 22.399 44.840 37.447 1.00 27.04 C

The floating point values in columns 6 through 8 are the coordinates of each atom.

The main website can be accessed at:

https://www.rcsb.org/pdb/home/home.do

Very briefly, the important keywords in a PDB file are as follows:

RECORD TITLE	DESCRIBES

HEADER	A BRIEF DESCRIPTION OF THE STRUCTURE, KIND OF A KEYWORD DESCRIBING THE BIOMOLECULE
TITLE	TITLE FOR THE EXPERIMENT OR ANALYSIS THAT IS REPRESENTED IN THE ENTRY
AUTHOR	PEOPLE WHO PERFORMED THE EXPERIMENT
REMARK	IS FOLLOWED BY 3, OR {0,1,2,4,5-299, 300-999}. PRESENTS EXPERIMENTAL DETAILS
SEQRES	LISTING OF CONSECUTIVE CHEMICAL COMPONENTS COVALENTLY LINKED TO FORM A POLYMER
ATOM	THESE RECORDS PRESENT ATOMIC COORDINATES FOR AMINO ACIDS AND NUCLEOTIDES (NATURAL)
HETATM	NON-POLYMER OR NON-STANDARD CHEMICAL COORDINATES

For a complete listing of the record description(s) can be had by referring to the following website:

http://www.wwpdb.org/documentation/file-format-content/format33/v3.3.html

SEQUENCE COMPARISON AND DATABASE SEARCH

Sequence comparison is undoubtedly the most basic operation in computational biology. It appears explicitly or implicitly in all subsequent chapters of this book. Let us start by understanding the following questions.

1. Why compare sequences?

2. Why do you need a computer for performing sequence comparisons?
3. How can you achieve it using computers?

Sequence comparison serves as a basis for many other, more complex, sequence manipulations. Roughly speaking, this operation consists of finding which parts of the sequences are alike and which parts differ. However, behind this apparently simple concept, a great variety of truly distinct problems exist, with diverse formalizations and sometimes requiring completely different data structures and algorithms for an efficient solution.

The use of computers hardly needs justification when we deal with large quantities of data, as in searching large databases in biology. Yet, even in cases where we could conceivably do comparisons "*by hand*" the use of computers is safer and more convenient. We will discuss some of the most commonly used methods for sequence comparison below. We will not go into the particulars of algorithms in a mathematical way, rather we would try to understand the concepts in a simpler "*lay man*" way.

PAIRWISE SEQUENCE COMPARISON

Comparing two sequences (protein or DNA) for similarities and differences is the most basic operation in computational molecular bioscience. The way we do this is called *alignment*. The two sequences are matched in an arbitrary way (*heuristic*), the quality of the match is scored, then one sequence is moved with respect to the other and the match is scored again. This is done till the best-score is found for a particular alignment. Although it sounds simple when explained this way, it is far more difficult to do this for sequences by hand and human eye cannot really distinguish between two optimal alignments. Hence, we need an automated way to accomplish this task. For this we need to ask a few questions and seek answers to them computationally.

- How should the alignments be scored?
- Which is the best scoring scheme?

- Should gaps be allowed for better matches elsewhere in the sequences?
- If gaps are allowed, then how should it be scored?
- Which algorithm is best for finding optimal alignment?
- Is the alignment significant?
- Given two random sequences, can you achieve an alignment of similar quality?

Several terms are commonly used when describing the alignment between two different sequences. The most frequently encountered ones are given below:

- *Sequence Identity*. Refers to the occurrence of the same nucleic acid or protein amino acid in the same position in the aligned sequences.
- *Sequence Similarity*. Substitutions of amino acids or nucleic acids are scored using scoring matrices. These matrices are constructed based on the probability with which such substitutions occur in nature. For example, amino acid residues of similar chemical properties are found to substitute for one another more readily than residues with different chemical properties.
- *Sequence Homology*. It indicates evolutionary relatedness between sequences.

Let us give some thought to scoring matrices and understand their role in sequence alignment.

SCORING MATRICES

What you really want to learn when evaluating a sequence alignment is whether a given alignment is *random*, or *meaningful*. If the alignment is meaningful, you want to gauge just how meaningful it is. You can do this by constructing a *scoring matrix*. A scoring matrix is a table of values that describe the probability of a residue pair (amino acid or base) occurring in an alignment. The values in a scoring matrix are logarithms of ratios of two probabilities.

- One is the probability of random occurrence of an amino acid in a sequence alignment. This value is simply the product of the independent frequencies of occurrence of each of the amino acids.
- The other is the probability of meaningful occurrence of a pair of residues in a sequence alignment.

These probabilities are derived from samples of actual sequence alignments that are known to be valid. In order to score an alignment, the alignment program needs to know if it is more likely that a given amino acid pair has occurred randomly, or that it has occurred as a result of an evolutionary event. The logarithm of the ratio of the probability of meaningful occurrence to the probability of random occurrence is positive if the probability of meaningful occurrence is greater, and negative if the probability of random occurrence is greater. Because the scores are logarithms of probability ratios, they can be meaningfully added to give a score for the entire sequence. The more positive the score, the more likely the alignment is to be significant.

BLOSUM matrices are derived from the *Blocks* database, a set of *ungapped* alignments of sequence regions from families of related proteins. A clustering approach sorts the sequences in each block into closely related groups, and the frequencies of substitutions between these within a family derives the probability of a meaningful substitution. The numerical value (e.g., 62) associated with a BLOSUM matrix represents the cut-off value for the clustering step. A value of 62 indicates that sequences were put into the same cluster if they were more than 62% identical. By allowing more diverse sequences to be included in each cluster, lower cut-off values represent longer evolutionary time scales, so matrices with low cut-off values are appropriate for seeking more distant relationships. BLOSUM62 is the standard matrix for ungapped alignments, while BLOSUM50 is more appropriate when generating gapped alignments. A sample BLOSUM62 matrix is given below.

	Ala	Arg	Asn	Asp	Cys	Gln	Glu	Gly	His	Ile	Leu	Lys	Met	Phe	Pro	Ser	Thr	Trp	Tyr	Val
Ala	4																			
Arg	-1	5																		
Asn	-2	0	6																	
Asp	-2	-2	1	6																

Cys	0	-3	-3	-3	9															
Gln	-1	1	0	0	-3	5														
Glu	-1	0	0	2	-4	2	5													
Gly	0	-2	0	-1	-3	-2	-2	6												
His	-2	0	1	-1	-3	0	0	-2	8											
Ile	-1	-3	=3	-3	-1	-3	-3	-4	-3	4										
Leu	-1	-2	-3	-4	-1	-2	-3	-4	-3	2	4									
Lys	-1	2	0	-1	-3	1	1	-2	-1	-3	-2	5								
Met	-1	-1	-2	-3	-1	0	-2	-3	-2	-1	2	-1	5							
Phe	-2	-3	-3	-3	-2	-3	-3	-3	-1	0	0	-3	0	6						
Pro	-1	-2	-2	-1	-3	-1	-1	-2	-2	-3	-3	-1	-2	-4	7					
Ser	1	-1	1	0	-1	0	0	0	-1	-2	-2	0	-1	-2	-1	4				
Thr	0	-1	0	-1	-1	-1	-1	-2	-2	-1	-1	-1	-1	-2	-1	1	5			
Trp	-3	-3	-4	-4	-2	-2	-3	-2	-2	-3	-2	-3	-1	1	-4	-3	-2	11		
Tyr	-2	-2	-2	-3	-2	-1	-3	2	-1	-1	-2	-1	3	-3	-2	-2	2	2	7	
Val	0	-3	-3	-3	-1	-2	-2	-3	-3	3	1	-2	1	-1	-2	-2	0	-3	-1	4

The other popular matrix is known as **PAM (Point Accepted Mutation)**; these are scaled according to a model of evolutionary distance from alignments of closely related sequences. One PAM *'unit'* is equivalent to an average change of 1% of all amino acid positions. The most commonly used PAM matrix is PAM250. We will not discuss PAM in this book.

Let us look at a simple pairwise sequence alignment. A typical pairwise alignment carried out using EMBOSS program '*needle*' is given below for comprehension. The EMBOSS suite can be installed or the programs in the suite can be accessed at the following website:

http://www.ebi.ac.uk/Tools/psa/emboss_needle/

A typical output looks like this:

```
########################################
# Program: needle
# Rundate: Thu 30 Mar 2017 08:03:02
# Commandline: needle
#    -auto
#    -stdout
#    -asequence emboss_needle-I20170330-080301-0281-52226113-pg.
```

```
asequence
#   -bsequence emboss_needle-I20170330-080301-0281-52226113-pg.
bsequence
#   -datafile EBLOSUM62
#   -gapopen 10.0
#   -gapextend 0.5
#   -endopen 10.0
#   -endextend 0.5
#   -aformat3 pair
#   -sprotein1
#   -sprotein2
# Align_format: pair
# Report_file: stdout
########################################

#=======================================
#
# Aligned_sequences: 2
# 1: ACD11_ARATH
# 2: SEQUENCE
# Matrix: EBLOSUM62
# Gap_penalty: 10.0
# Extend_penalty: 0.5
#
# Length: 207
# Identity:     206/207 (99.5%)
# Similarity:   206/207 (99.5%)
# Gaps:           1/207 ( 0.5%)
# Score: 1044.0
#
```

```
#
#=======================================

ACD11_ARATH       1
-MADSEADKPLRKISAAFKKLAIIVNSPNPEVPVTQFSHACS
LVSPLFGC    49
             ||||||||||||||||||||||||||||||||||||||||||||||||||
SEQUENCE         1
SMADSEADKPLRKISAAFKKLAIIVNSPNPEVPVTQFSHACSL
VSPLFGC    50

ACD11_ARATH      50 LGIAFKFAEMDYVAKVDDLVRA
SSSISTLVVMM DKDIEADCVRKAGSHTR    99
             ||||||||||||||||||||||||||||||||||||||||||||||||||
SEQUENCE        51 LGIAFKFAEMDYVAKVDD LVRASSSISTLVV
MMDKDIEADCVRKAGSHTR   100

ACD11_ARATH     100 NLLRVKRGLDMVKVLF
EQIIASEGDNSLKDP ATKSYAQVFAPHHGWAIRK    149
             ||||||||||||||||||||||||||||||||||||||||||||||||||
SEQUENCE       101 NLLRVKRGLD MVKVLFEQIIASEGDNSL
KDPATKSYAQVFAPHHGWAIRK   150

ACD11_ARATH     150 AVSLGMYALPTRAHLLNMLKEDEAAAKI
HMQSYVNSSAPLITYLDNLFLS   199
             ||||||||||||||||||||||||||||||||||||||||||||||||||
SEQUENCE       151 AVSLGMYALPTRAHLLNMLKEDEAAAKI
HMQSYVNSSAPLITYLDNLFLS   200

ACD11_ARATH     200 KQLGIDW   206
             |||||||
SEQUENCE       201 KQLGIDW   207
```

#-------------------------------------

#-------------------------------------

In the above case, we have two sequences, **O64587** and PDB code **4NT1**, from *Arabidopsis thaliana*. One is sequence from genomics experiments with protein level evidence, whereas the other is the sequence for which the crystal structure was solved. Except for the N-terminal **Ser** rest of the residues are identical. In practice, you will encounter this situation less often. Now, we will be discussing gaps in alignments.

Gaps

Change is the law of nature. At a molecular level biomolecular sequences also go changes in their basic composition. These changes in amino acid residues of proteins or DNA bases are called *mutations*. The most frequent change is that of a single residue being replaced with another naturally occurring residue is known as *point mutation*. Point mutations are the most common way a protein evolves. The functional implications of such mutations are manifold, sometimes beneficial and sometimes deleterious. Insertions and deletions of residues are the other ways of changes in protein or DNA sequences. Hence, it is necessary to introduce gaps into one or both of the sequences being aligned to produce a meaningful alignment between them.

Most algorithms use a *gap penalty* to represent the validity of adding a gap in an alignment. The addition of a gap comes at a cost. This is discussed in terms of the *overall alignment score*, wherein gaps will open only where they are really needed and not all over the sequence. Most sequence alignment models use *affine gap penalties*. In these schemes, the cost of *opening* a gap in a sequence is different from the cost of *extending* an already opened gap. Of these two penalties—-the gap opening penalty and the gap extension penalty—-the gap opening penalties tend to be much costlier (higher) than the associated extension penalty. This tendency reflects insertions and deletions to occur over several residues at a time. Gap penalties are intimately tied to the scoring matrix that aligns the sequences: the best pair of gap opening and extension penalties for one scoring matrix doesn't necessarily work with another. Scores of -11 for gap opening and -1 for gap extension are commonly used in conjunction with the BLOSUM62 matrix for gapped-BLAST, while BLOSUM50 uses a -12/-1 penalty. You may notice that in the previous example of global alignment, the penalties have been set to -10/-0.5. Can you think why was this done?

Global alignments are done along the whole length of the two sequences. The algorithm is called **Needleman-Wunsch**, and appropriately called '*needle*' in EMBOSS suite. In the language of programming this belongs to the class of *Dynamic Programming*. It was introduced by Saul Needleman and Christian Wunsch for biological sequence comparison. It essentially solves an *optimization* problem, with a goal of maximizing the total score for the alignment. We will not go into the details of the algorithm here.

Local Alignment

This is the most commonly used strategy for sequence alignment. In global alignment, the two sequences are assumed to be known and are aligned over full length. In practice, you would be searching your sequence (query) against a database of sequences (NCBI, UniProt etc). A different version of dynamic programming that performs local alignment of two sequences is called **Smith-Waterman** algorithm, after its inventors Dr. Temple Smith and Dr. Michael Waterman. In this, alignment isn't required to extend from beginning to end of the two sequences being aligned. If the cumulative score, up to some point, in the sequence is negative the alignment can be abandoned and a new alignment started. The alignment can also end anywhere in the matrix.

BLAST is the preferred algorithm for performing local alignments. It does alignments in three steps:

- It creates a list of all short sequences (called *words*) which score above a threshold value when aligned with query sequence. Typical word lengths are 3 (proteins) and 11 (nucleic acids).
- The sequence database is searched for occurrences of these words. Matching words are then extended into ungapped local alignments. Extensions are continued till the score drops below a threshold.
- Top scoring alignments in a sequence (or maximal-scoring segment pairs (MSPs)) are combined into local alignments.

Gapped alignments are now possible with BLAST.

BLAST PROGRAM

There are many ways to use BLAST. Many online sequence databases provide BLAST on their website. One can download and install BLAST binaries for offline use. It comes with four main executables.

- *blastall* - performs BLAST searches using one of five BLAST programs: blastp, blastn, blastx, tblastn, or tblastx.
- *blastpgp* - Performs searches in PSI-BLAST or PHI-BLAST mode.
- *bl2seq* - Performs a local alignment of two sequences.
- *formatdb* - Converts a FASTA-format flat file sequence database into a BLAST database.

blastall encompasses all the major options for ungapped and gapped BLAST searches. A full list of its command-line arguments can be displayed with the command *blastall* -

[-p]

Program name. Its options include:

blastp

Protein sequence (PS) query versus PS database

blastn

Nucleic acid sequence (NS) query versus NS database

blastx

NS query translated in all six reading frames versus PS database

tblastn

PS query versus NS database dynamically translated in all six reading frames

tblastx

Translated NS query versus translated NS database—computationally intensive

[-d]

Database name. Each indexed database consists of several files; the name is the common portion of those filenames.

[-i]

Query sequence filename. Defaults to standard input if not specified.

[-e]

Expectation value cutoff. (See Section 7.8.1.5.)

[-m]

Alignment view. Its options include:

0

Pairwise

1

Master-slave, show identities

2

Master-slave, no identities

3

Flat master-slave, show identities

4

Flat master-slave, no identities

5

Master-slave, no identities, blunt ends

6

Flat master-slave, no identities, blunt ends

[-o]

Output file name. Defaults to standard output if not specified.

[-G]

Gap opening penalty. Defaults to 11 (for BLOSUM63 matrix).

[-E]

Gap extension penalty. Defaults to 1 (for BLOSUM63 matrix).

[-q]

Nucleotide mismatch penalty. *blastn* only. Defaults to -3.

[-r]

Reward for nucleotide match. *blastn* only. Defaults to 1.

[-b]

Number of alignments to show.

[-g]

Perform gapped alignment. T/F. Unavailable with *tblastx*.

[-M]

Scoring matrix. Defaults to BLOSUM62.

[-T]

Produce HTML output. T/F. Defaults to F.

Many other options are available, but these are the most commonly used ones.

blastpgp allows you to use two new BLAST modes: PHI-BLAST (Pattern Hit Initiated BLAST) and PSI-BLAST (Position Specific Iterative BLAST). PHI-BLAST uses protein motifs, such as those found in PROSITE and other motif databases, to increase the likelihood of finding biologically significant matches. PSI-BLAST uses an iterative alignment procedure to develop position-specific scoring matrices, which increases its capability to detect weak pattern matches.

bl2seq allows the comparison of two known sequences using the *blastp* or *blastn* programs. Most of the command-line options for *bl2seq* are similar to those for *blastall.*

INTEGRATED TOOL FOR SEQUENCE ANALYSIS

Most of the sequence related changes can be addressed by a fantastic suite of programs made available by **Sanger Institute**. I highly recommend getting familiar with all the features of the **EMBOSS** package, briefly discussed below.

Emboss

EMBOSS stands for (European Molecular Biology Open Software Suite) is an open source package of sequence analysis tools. This software covers a wide range of functionality and can handle data in a variety of formats. Extensive libraries are provided with the package, allowing users to develop and release their own software. EMBOSS also integrates a range of currently available packages and tools for sequence analysis, such as BLAST and ClustalW. A Java API (*Jemboss*) is also available. EMBOSS contains around 150 programs (*applications*). These are just some of the areas covered:

- Sequence alignment
- Rapid database searching with sequence patterns
- Protein motif identification, including domain analysis
- Nucleotide sequence pattern analysis, for example to identify CpG islands or repeats
- Codon usage analysis for small genomes
- Rapid identification of sequence patterns in large scale sequence sets
- Presentation tools for publication

...and much more.

EMBOSS programs have common functionality. All of them understand different sequence formats, output formats, and feature formats. We take up a few themes common to majority of the EMBOSS programs:

UNIFORM SEQUENCE ADDRESS (USA)

The USA is a standard sequence naming used by all EMBOSS applications. The syntax for usage is as follows:

- format::file
- format::file:entry
- dbname:entry
- @listfile (a list of filenames)

The '::' and ':' syntax is to allow both database names and sequence format (e.g., *'embl'* or *'pir'* etc.). Subsequences can be specified after the main body of the USA either as [*start*: *end*] or [*start* : *end* : r]. Please refer to documentation for more command line options.

SEQUENCE FORMATS

EMBOSS accepts the following sequence formats.

abi, acedb, clustal, aln, codata, dbid, em, pearson, gcg, gcg8, genbank, gb, ddbj, gff, henning86, ig, jackknifer, nbrf, pir, nexus, paup, nexusnon/paupnon, treecon, mega, meganon, msf, ncbi, pfam, stockholm, phylip, selex, staden, experiment, strider, swissprot, swiss, sw, text, plain, raw, asis.

OUTPUT SEQUENCE FORMATS

It can output in the above listed formats, as well as some more.

ALIGNMENT FORMATS

Multiple sequence alignment formats and pairwise sequence alignment formats are also supported.

EMBOSS APPLICATION GROUPS

EMBOSS developers have clustered the programs into application groups. These groups are given below for easy reference and further exploration.

Alignment consensus, Alignment differences, Alignment dot plots, Alignment global, Alignment global, Alignment local, Alignment multiple, Display, Edit, Enzyme kinetics, Feature Tables, Information, Menus, Nuclei 2d structure, Nucleic codon usage, Nucleic composition, Nucleic vpg islands, Nucleic gene finding, Nucleic motifs, Nucleic mutation, Nucleic primers, Nucleic profiles, Nucleic repeats, Nucleic restriction, Nucleic transcription, Nucleic translation, Phylogeny, Protein 2d structure, Protein 3d structure, Protein composition, Protein motifs, Protein mutation, Protein profiles, Protein structure, Test and Utilities for database creation, indexing, miscellaneous.

One can consult the online documentation of EMBOSS suite for a deeper understanding of the programs which are grouped under the above applications. This would be really helpful, if the objectives of the study are clear in mind.

REFERENCES

1. Rice P, Longden I, Bleasby A. "EMBOSS: the European Molecular Biology Open Software Suite" *Trends in Genetics* **16** (6): 276-277 (**2000**).
2. Altschul S, Warren G, Miller Webb, Myers E, Lipman D. "Basic Local Alignment Search Tool" *J. Mol. Biol.* **215** (3): 403-410 (**1990**).
3. Lipman DJ, Pearson WR. "Rapid and Sensitive Protein Similarity Searches" *Science* **227** (4693): 1435-1441 (**1985**)
4. Berman HM, Westbrook J, Feng Z, Gililand G, Bhat TN, Weissig H, Shindyalov IN, Bourne PE. "The Protein Data Bank" *Nucleic Acids Research*, **28**: 235-242 (**2008**)

4

R IN BIOINFORMATICS

This introductory chapter is intended for you to get familiar with the statistical package R. R (www.r-project.org) is a commonly used open source statistical software with wide following and regular updates. **R** allows you to carry out statistical analyses in an interactive mode, as well as allowing simple to complex programming. We will learn the computational basics in this chapter and see the applications in the subsequent chapters. Let us start with installing the software on your computers.

INSTALLING R

To use R, you first need to install the R program on your computer.

The first thing to do is to check whether R is already installed on your computer (for example, by a previous user).

If you are using a Windows PC, there are two ways you can check whether R is already installed on your computer:

- Check if there is an "R" icon on the desktop of the computer that you are using. If so, double-click on the "R" icon to start R. If you cannot find an "R" icon, try step 2 instead.

- Click on the "Start" menu at the bottom left of your Windows desktop, and then move your mouse over "All Programs" in the menu that pops up. See if "R" appears in the list of programs that pops up. If it does, it means that R is already installed on your computer, and you can start R by selecting "R" (or *R X.X.X*, where *X.X.X* gives the version of R, e.g., *R 3.3.1*) from the list.

If either (1) or (2) above does succeed in starting R, it means that R is already installed on the computer that you are using. (If neither succeeds, R is not installed yet). If there is an old version of R installed on the Windows PC that you are using, it is worth installing the latest version of R, to make sure that you have all the latest R functions available to you to use.

These instructions will focus on installing R on a Windows PC. However, I will also briefly mention how to install R on a Macintosh or Linux computer (see below).

FINDING OUT WHAT IS THE LATEST VERSION OF R

To find out what is the latest version of R, you can look at the CRAN (Comprehensive R Network) website

https://cran.r-project.org/

You can find the latest release besides "The latest release" (about half way down the page). It will show something like "R-X.X.X.tar.gz" (e.g., "R-3.3.1.tar.gz"). This means that the latest release of R is X.X.X (for example, 3.3.1).

New releases of R are made very regularly (approximately once a month), as R is actively being improved all the time. It is worthwhile installing new versions of R regularly. Working the latest version of R is recommended for short-term projects (to ensure compatibility with all the latest versions of the R packages that you have downloaded).

A caution for those involved in long-term investigations. Try to maintain version consistency. Hence, complete your projects using a single version of **R**. Upon switching to a newer version of R, please make sure that the results are *reproducible* in its entirety.

INSTALLING R ON A WINDOWS PC

To install R on your Windows computer, follow these steps:

1. Go to *http://ftp.heanet.ie/mirrors/cran.r-project.org*
2. Under "Download and Install R", click on the "Windows" link.
3. Under "Subdirectories", click on the "base" link.
4. On the next page, you should see a link with something like "Download R 3.3.1 for Windows" (or R X.X.X, where X.X.X gives the version of R, e.g., R 3.3.1). Click on this link.
5. You may be asked if you want to save or run a file "R-3.3.1-win64.exe." Choose "Save" and save the file on the Desktop. Then double-click on the icon for the file to run it.
6. You will be asked what language to install it in - choose English.
7. The R Setup Wizard will appear in a window. Click "Next" at the bottom of the R Setup wizard window.
8. The next page says "Information" at the top. Click "Next" again.
9. The next page says "Information" at the top. Click "Next" again.
10. The next page says "Select Destination Location" at the top. By default, it will suggest to install R in "C:\Program Files" on your computer.
11. Click "Next" at the bottom of the R Setup wizard window.
12. The next page says "Select components" at the top. Click "Next" again.
13. The next page says "Startup options" at the top. Click "Next" again.
14. The next page says "Select start menu folder" at the top. Click "Next" again.

15. The next page says "Select additional tasks" at the top. Click "Next" again.

16. R should now be installed. This will take about a minute. When R has finished, you will see "Completing the R for Windows Setup Wizard" appear. Click "Finish".

17. To start R, you can either follow step 18, or 19:

18. Check if there is an "R" icon on the desktop of the computer that you are using. If so, double-click on the "R" icon to start R. If you cannot find an "R" icon, try step 19 instead.

19. Click on the "Start" button at the bottom left of your computer screen, and then choose "All programs", and start R by selecting "R" (or R X.X.X, where X.X.X gives the version of R, e.g., R 3.3.1) from the menu of programs.

20. The R console (a rectangle) should pop up:

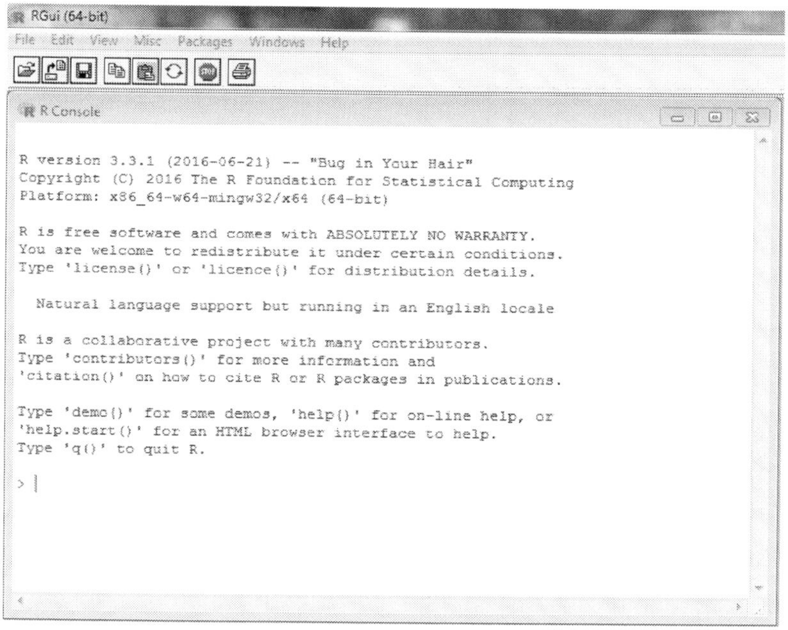

HOW TO INSTALL R ON NON-WINDOWS COMPUTERS

The instructions above are for installing R on a Windows PC. If you want to install R on a computer that has a non-Windows operating system (Macintosh or Linux), you should download the appropriate R installer for that operating system at

http://ftp.heanet.ie/mirrors/cran.r-project.org

and follow the R installation instructions for the appropriate operating system at

http://ftp.heanet.ie/mirrors/cran.r-project.org/doc/FAQ/R-FAQ.html#How-can-R-be-installed_003f

INSTALLING R PACKAGES

R comes with some standard packages that are installed when you install **R**. In this chapter we will also tell you how to use **R** packages that are useful in biomolecular science, for example, the "*rmeta*" package. These additional packages do not come with the standard installation of **R**. Hence, you need to install them by yourself.

HOW TO INSTALL AN R PACKAGE

Once you have installed **R** on a computer, you can install an additional package by following the steps below:

1. To start **R**, follow either step 2 or 3:
2. Check if there is an "**R**" icon on the desktop of the computer that you are using. If so, double-click on the "**R**" icon to start **R**. If you cannot find an "**R**" icon, try step 3 instead.
3. Click on the "Start" button at the bottom left of your computer screen, and then choose "All programs", and start **R** by selecting "**R**" (or *R X.X.X*, where *X.X.X* gives the version of **R**, e.g., *R 3.3.1*) from the menu of programs.
4. The **R** console (a rectangle) should pop up.

5. Once you have started **R**, you can now install an **R** package (e.g., the "*rmeta*" package) by choosing "Install package(s)" from the "Packages" menu at the top of the **R** console. This will ask you what website you want to download the package from, you should choose "USA" (or another country, if you prefer). It will also bring up a list of available packages that you can install, and you should choose the package that you want to install from that list (e.g., "*rmeta*").
6. This will install the "*rmeta*" package.
7. The "*rmeta*" package is now installed. Whenever you want to use the "*rmeta*" package after this, after starting **R**, you first have to load the package by typing into the **R** console:

```
> library("rmeta")
```

Note that there are some additional **R** packages for bioinformatics that are part of a special set of **R** packages called **Bioconductor** (www.bioconductor.org) such as the "*yeastExpData*" **R** package, the "*Biostrings*" **R** package, etc.). These **Bioconductor** packages need to be installed using a different, **Bioconductor**-specific procedure (see How to install a **Bioconductor R** package below).

HOW TO INSTALL A BIOCONDUCTOR R PACKAGE

The procedure above can be used to install the majority of **R** packages. However, the **Bioconductor** set of bioinformatics **R** packages need to be installed by a special procedure. **Bioconductor** (*www.bioconductor.org*) is a group of **R** packages that have been developed for bioinformatics. This includes R packages such as "*yeastExpData*", "*Biostrings*", etc.

To install the **Bioconductor** packages, follow these steps:

- To start R, follow either step 2 or 3:
- Check if there is an "**R**" icon on the desktop of the computer that you are using. If so, double-click on the "**R**" icon to start **R**. If you cannot find an "**R**" icon, try step 3 instead.
- Click on the "Start" button at the bottom left of your computer screen, and then choose "All programs", and start **R** by

selecting "**R**" (or *R X.X.X*, where *X.X.X* gives the version of **R**, e.g., *R 3.3.1*) from the menu of programs.

- The **R** console (a rectangle) should pop up.
- Once you have started **R**, now type in the **R** console:

```
> source("http://bioconductor.org/biocLite.R")
> biocLite()
```

- This will install a core set of **Bioconductor** packages *("affy", "affydata", "affyPLM", "annaffy", "annotate", "Biobase", "Biostrings", "DynDoc", "gcrma", "genefilter", "geneplotter", "hgu95av2.db", "limma", "marray", "matchprobes", "multtest", "ROC", "vsn", "xtable", "affyQCReport")*. This takes a few minutes.
- At a later date, you may wish to install some extra Bioconductor packages that do not belong to the core set of Bioconductor packages. For example, to install the Bioconductor package called "*yeastExpData*", start **R** and type in the **R** console:

```
> source("http://bioconductor.org/biocLite.R")
> biocLite("yeastExpData")
```

- Whenever you want to use a package after installing it, you need to load it into **R** by typing:

```
> library("yeastExpData")
```

RUNNING R

To start **R**, you can either follow step 1 or 2:

1. Check if there is an "**R**" icon on the desktop of the computer that you are using. If so, double-click on the "**R**" icon to start **R**. If you cannot find an "**R**" icon, try step 2 instead.
2. Click on the "Start" button at the bottom left of your computer screen, and then choose "All programs", and start **R** by selecting "**R**" (or *R X.X.X*, where *X.X.X* gives the version of **R**, e.g., *R 3.3.1*) from the menu of programs.

This should bring up a new window, which is the *R console*.

RSTUDIO

To readers who are used to programming in Integrated Development Environment (IDE), *RStudio* is the best option. *RStudio* is a free open-source IDE for **R**. It was founded by J.J. Allaire and Hadley Wickham is the Chief Scientist at *RStudio*. It is available in two versions.

- *RStudio Desktop*: program is run locally on a desktop computer or laptop.
- *RStudio Server*: allows accessing *RStudio* using a web browser while it is running on a remote Linux server.

It is also available as a commercial edition. Please follow the installation instructions at:

www.rstudio.com

This is an alternative to working on **R** console. RStudio requires prior installation of **R**.

We start with a brief introduction to R, with hands-on training of basic mathematical and statistical operations in R.

A BRIEF INTRODUCTION TO R

You will type **R** commands into the **R** console in order to carry out analyses in **R**. In the **R** console you will see:

```
>
```

This is the **R** prompt. We type the commands needed for a particular task after this prompt. The command is carried out after you hit the Return key. Once you have started **R**, you can start typing in commands, and the results will be calculated immediately. Try the following.

```
> 3*3
[1] 9
> 10-4
[1] 6
```

All variables (scalars, vectors, matrices, etc.) created by **R** are called *objects*. In **R**, we assign values to variables using an arrow. For example, we can assign the value 3*3 to the variable *x* using the command:

```
> x <- 3*3
```

To view the contents of any **R** object, just type its name, and the contents of that **R** object will be displayed:

```
> x
[1] 6
```

There are several possible different types of objects in **R**, including *scalars*, *vectors*, *matrices*, *arrays*, *dataframes*, *tables*, and *lists*. The scalar variable *x* above is one example of an **R** object. While a scalar variable such as *x* has just one element, a vector consists of several elements. The elements in a vector are all of the same type (e.g., *numeric* or *characters*), while lists may include elements such as characters as well as numeric quantities.

To create a vector, we can use the *c()* (combine) function. For example, to create a vector called *vector_1* that has elements with values 10, 6, 9 and 5, we type:

```
> vector_1 <- c(10, 6, 9, 5)
```

To see the contents of the variable *vector_1*, we can just type its name:

```
> vector_1
[1]  10  6  9  5
```

The [1] is the index of the first element in the vector. We can extract any element of the vector by typing the vector name with the index of that element given in square brackets. For example, to get the value of the 2nd element in the vector *vector_1*, we type:

```
> vector_1[2]
[1] 6
```

In contrast to a *vector*, a *list* can contain elements of different types, for example, both numeric and character elements. A list can also include other variables such as a vector. The *list()* function is used to create a list. Let us create a list *list* _1 as follows:

```
> list_1 <- list(name="Deepak", wife="Shalini", vector_1)
```

We can then print out the contents of the list *list_1* by typing its name:

```
> list_1
$name
```

[1] "Deepak"

$wife

[1] "Shalini"

[[3]]

[1] 10 6 9 5

The elements in a list are numbered, and can be referred to using indices. We can extract an element of a list by typing the list name with the index of the element given in double square brackets (in contrast to a vector, where we only use single square brackets). Thus, we can extract the second and third elements from *list _1* by typing:

> list_1[[2]]

[1] "Shalini"

> list_1[[3]]

[1] 10 6 9 5

Elements of lists may also be named, and in this case the elements may be referred to by giving the list name, followed by "$", followed by the element name. For example, *list_1$name* is the same as *list_1[[1]]* and *list_1$wife* is the same as *list_1[[2]]*:

> list_1$wife

[1] "Shalini"

We can find out the names of the named elements in a list by using the *attributes ()* function, for example:

> attributes(list_1)

$names

[1] "name" "wife" ""

When you use the *attributes ()* function to find the named elements of a list variable, the named elements are always listed under a heading "$names." Therefore, we see that the named elements of the list variable *list_1* are called "name" and "wife", and we can retrieve their values by typing *list_1$name* and *list_1$wife*, respectively.

Another type of object that you will encounter in **R** is a *table* variable. For example, if we made a vector variable *mynames* containing the names of children in a class, we can use the *table ()* function to produce a table variable that contains the number of children with each possible name:

> mynames <- c("Mary", "John", "Ann", "Sinead", "Joe", "Mary", "Jim", "John", "Simon")

> table(mynames)

mynames

```
 Ann  Jim  Joe  John  Mary Simon Sinead
  1    1    1    2     2    1     1
```

We can store the table variable produced by the function *table ()*, and call the stored table "mytable", by typing:

> mytable <- table(mynames)

To access elements in a table variable, you need to use double square brackets, just like accessing elements in a list. For example, to access the fourth element in the table *mytable* (the number of children called "John"), we type:

> mytable[[4]]

[1] 2

Alternatively, you can use the name of the fourth element in the table ("John") to find the value of that table element:

> mytable[["John"]]

[1] 2

Functions in **R** usually require *arguments*, which are input variables (i.e., objects) that are passed to them, which they then carry out some operation on. For example, the *log10 ()* function is passed a number, and it then calculates the *log* to the *base 10* of that number:

> log10(100)

2

In **R**, you can get help about a particular function by using the *help ()* function. For example, if you want help about the *log10 ()* function, you can type:

> help("log10")

When you use the *help ()* function, a box or webpage will pop up with information about the function that you asked for help with.

If you are not sure of the name of a function, but think you know part of its name, you can search for the function name using the *help. Search ()* and *RSiteSearch()* functions. The *help.search()* function searches to see if you already have a function installed (from one of the **R** packages that you have installed) that may be related to some topic you're interested in. The *RSiteSearch()* function searches all **R** functions (including those in packages that you haven't yet installed) for functions related to the topic you are interested in.

For example, if you want to know if there is a function to calculate the standard deviation of a set of numbers, you can search for the names of all installed functions containing the word "deviation" in their description by typing:

> help.search("deviation")

Help files with alias or concept or title matching 'deviation' using fuzzy matching:

genefilter::rowSds

Row variance and standard deviation of a numeric array

nlme::pooledSD Extract Pooled Standard Deviation

stats::mad Median Absolute Deviation

stats::sd Standard Deviation

vsn::meanSdPlot Plot row standard deviations versus row

Among the functions that were found, is the function *sd()* in the "*stats*" package (an **R** package that comes with the standard **R** installation), which is used for calculating the standard deviation.

In the example above, the *help.search()* function found a relevant function (*sd()* here). However, if you did not find what you were looking for with *help.search()*, you could then use the *RSiteSearch()* function to see if a search of all functions described on the **R** website may find something relevant to the topic that you're interested in:

> RSiteSearch("deviation")

The results of the *RSiteSearch()* function will be hits to descriptions of **R** functions, as well as to **R** mailing list discussions of those functions.

We can perform computations with **R** using objects such as scalars and vectors. For example, to calculate the average of the values in the vector *myvector* (i.e., the average of 10, 6, 9 and 5), we can use the *mean()* function:

```
> mean(myvector)
[1] 7.5
```

We have been using built-in **R** functions such as *mean()*, *length()*, print(), *plot()* etc. We can also create our own functions in **R** to do calculations that you want to carry out very often on different input data sets. For example, we can create a function to calculate the value of 20 plus square of some input number:

```
> myfunction <- function(x) { return(20 + (x*x)) }
```

This function will calculate the square of a number (*x*), and then add 20 to that value. The *return()* statement returns the calculated value. Once you have typed in this function, the function is then available for use. For example, we can use the function for different input numbers (e.g., 10, 25):

```
> myfunction(10)
[1] 120
> myfunction(25)
[1] 645
```

To quit R, type:

```
> q()
```

BIOCONDUCTOR AND SEQINR

Many authors have written R packages for performing a wide variety of analyses. These do not come with the standard R installation, but must be installed and loaded as "*add-ons*." Bioinformaticians have written several specialised *packages* for R. In this section, you will learn to use the SeqinR package to retrieve sequences from a DNA sequence database, and to carry out simple analyses of DNA sequences. Some well-known bioinformatics packages for R are the Bioconductor set of R packages (*www.bioconductor.org*), which contains several packages with many R functions for analysing biological data sets such as microarray data; and the SeqinR package (*pbil.univ-lyon1.fr/software/seqinr/home.php?lang=eng*), which contains R functions for obtaining sequences from DNA and protein sequence databases, and for analysing DNA and protein sequences.

To use function from the SeqinR package, we first need to install the SeqinR package (for instructions on how to install an R package, see above). Once you have installed the "SeqinR" R package, you can load the "SeqinR" R package by typing:

> library("seqinr")

Remember that you can ask for more information about a particular **R** command by using the *help()* function. For example, to ask for more information about the *library()* function, you can type:

> help("library")

THE NCBI SEQUENCE DATABASE

We have already seen what NCBI stands for and biological sequence data is stored in it. We have also seen how to query the NCBI database for information. The National Centre for Biotechnology Information (NCBI) (*www.ncbi.nlm.nih.gov*) in the US maintains a huge database of all the DNA and protein sequence data that has been collected, the NCBI Sequence Database. This also a similar database in Europe, the European Molecular Biology Laboratory (EMBL) Sequence Database (*www.ebi.ac.uk/embl*), and also a similar database in Japan, the DNA Data Bank of Japan (DDBJ; *www.ddbj.nig.ac.jp*). These three databases exchange data every night, so at any one point in time, they contain almost identical data. Interested readers are requested to delve into these resources.

Let us get into the details of how sequences are stored in NCBI. Each sequence in the NCBI Sequence Database is stored in a separate *record*, and is assigned a unique identifier that can be used to refer to that sequence record. The identifier is known as an *accession*, and consists of a mixture of numbers and letters. For example, Dengue virus causes Dengue fever, which is classified as a neglected tropical disease by the WHO. by any one of four types of Dengue virus: DEN-1, DEN-2, DEN-3, and DEN-4. The NCBI accessions for the DNA sequences of the

DEN-1, DEN-2, DEN-3, and DEN-4 Dengue viruses are NC_001477, NC_001474, NC_001475 and NC_002640, respectively.

Note that because the NCBI Sequence Database, the EMBL Sequence Database, and DDBJ exchange data every night, the DEN-1 (and DEN-2, DEN-3, DEN-4) Dengue virus sequence will be present in all three databases, but it will have different accessions in each database, as they each use their own numbering systems for referring to their own sequence records.

RETRIEVING GENOME SEQUENCE DATA VIA THE NCBI WEBSITE

You can easily retrieve DNA or protein sequence data from the NCBI Sequence Database *via* its website.

The Dengue DEN-1 DNA sequence is a viral DNA sequence, and as mentioned above, its NCBI accession is NC_001477. To retrieve the DNA sequence for the Dengue DEN-1 virus from NCBI, go to the NCBI website, type "NC_001477" in the Search box at the top of the webpage, and press the "Search" button beside the Search box:

On the results page you will see the number of hits to "NC_001477" in each of the NCBI databases on the NCBI website. There are many databases on the NCBI website, for example, the "*PubMed*" data contains abstracts from scientific papers, the "*Nucleotide*" database contains DNA and RNA sequence data, the "*Protein*" data contains protein sequence data, and so on. The picture below shows what the results page should look like for your NC_001477 search. As you are looking for the DNA sequence of the Dengue DEN-1 virus genome, you expect to see a hit in the NCBI Nucleotide database, and indeed there is a hit in the Nucleotide database (indicated by the "1" beside the icon for the Nucleotide database):

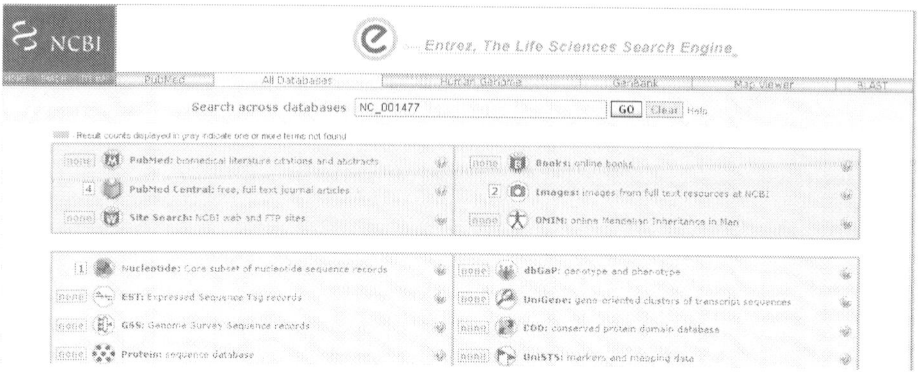

To look at the one sequence found in the Nucleotide database, you need to click on the icon for the NCBI Nucleotide database on the results page for the search:

When you click on the icon for the NCBI Nucleotide database, it will bring you to the record for NC_001477 in the NCBI Nucleotide database. This will contain the name and NCBI accession of the sequence, as well as other details such as any papers describing the sequence:

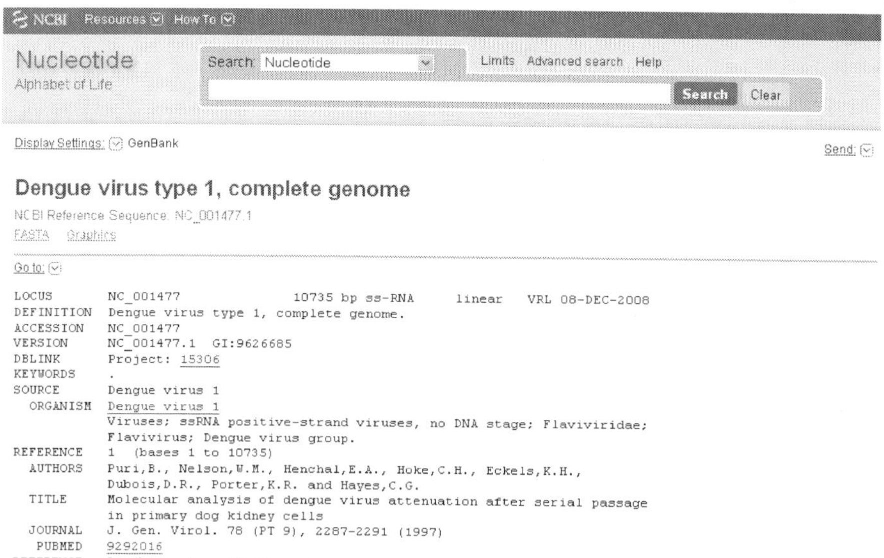

To retrieve the DNA sequence for the DEN-1 Dengue virus genome sequence as a FASTA format sequence file, click on "*Send*" at the top right of the NC_001477 sequence record webpage, and then choose "*File*" in the pop-up menu that appears, and then choose FASTA from the "Format" menu that appears, and click on "*Create file*".

A box will pop up asking you what to name the file, and where to save it. You should give it a sensible name (e.g., "den1.fasta") and save it in a place where you will remember (e.g., in the "My Documents" folder is a good idea):

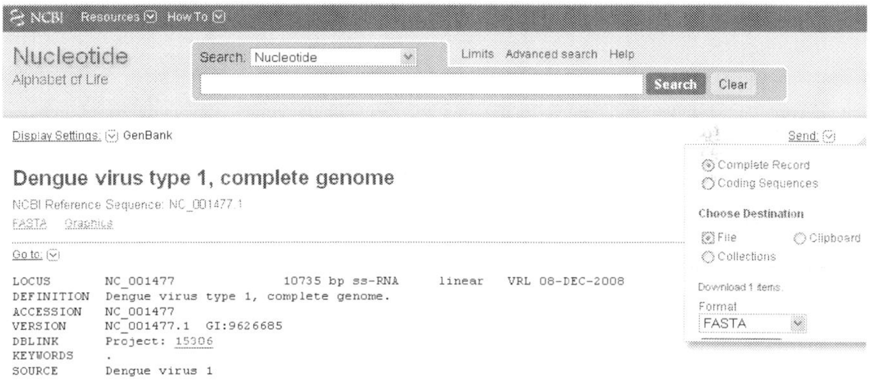

You can now open the FASTA file containing the DEN-1 Dengue virus genome sequence using WordPad on your computer. To open WordPad, click on "Start" on the bottom left of your screen, click on "All Programs" in the menu that appears, and then select "Accessories" from the menu that appears next, and then select "WordPad" from the menu that appears next. WordPad should start up. In Wordpad, choose "Open" from the "File" menu. The WordPad "Open" dialog will appear. Set "Files of type" to "All Documents" at the bottom of the WordPad "Open" dialog. You should see a list of files, now select the file that contains the DEN-1 Dengue virus sequence (e.g., "den1.fasta"). The contents of the FASTA format file containing the Dengue DEN-1 sequence should now be displayed in WordPad:

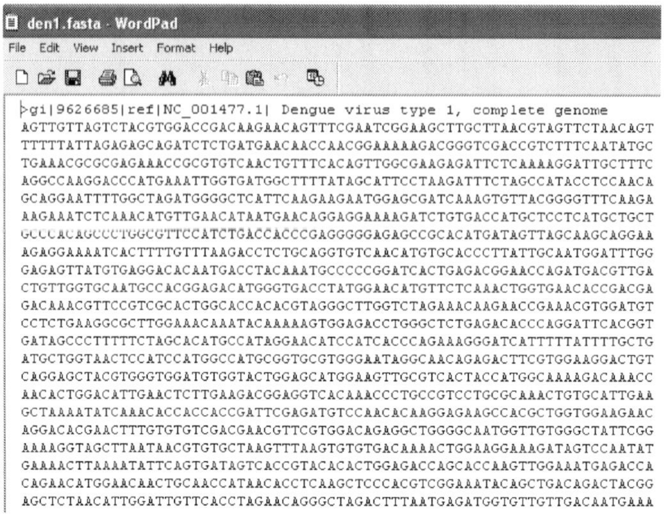

RETRIEVING GENOME SEQUENCE DATA USING SEQINR

Instead of going to the NCBI website to retrieve sequence data from the NCBI database, you can retrieve sequence data from NCBI directly from **R**, by using the **SeqinR R** package. For example, you learnt above how to retrieve the DEN-1 Dengue virus genome sequence, which has NCBI accession NC_001477, from the NCBI website. To retrieve a sequence with a particular NCBI accession, you can use **R** function "*getncbiseq()*" below, which you will first need to copy and paste into R:

```
> getncbiseq <- function(accession)
  {
    require("seqinr") # this function requires the SeqinR R package
    # first find which ACNUC database the accession is stored in:
    dbs <- c("genbank","refseq","refseqViruses","bacterial")
    numdbs <- length(dbs)
    for (i in 1:numdbs)
    {
      db <- dbs[i]
```

```
    choosebank(db)
    # check if the sequence is in ACNUC database 'db':
    resquery <- try(query(".tmpquery", paste("AC=", accession)), silent
= TRUE)
    if (!(inherits(resquery, "try-error")))
    {
      queryname <- "query2"
      thequery <- paste("AC=",accession,sep="")
      query(`queryname`,`thequery`)
      # see if a sequence was retrieved:
      seq <- getSequence(query2$req[[1]])
      closebank()
      return(seq)
    }
    closebank()
  }
  print(paste("ERROR: accession",accession,"was not found"))
}
```

Once you have copied and pasted the function *getncbiseq()* into **R**, you can use it to retrieve a sequence from the NCBI Nucleotide database, such as the sequence for the DEN-1 Dengue virus (accession NC_001477):

```
> dengueseq <- getncbiseq("NC_001477")
```

The variable *dengueseq* is a vector containing the nucleotide sequence. Each element of the vector contains one nucleotide of the sequence. Therefore, to print out a certain subsequence of the sequence, we just need to type the name of the vector *dengueseq* followed by the square brackets containing the indices for those nucleotides. For example, the following command prints out the first 50 nucleotides of the DEN-1 Dengue virus genome sequence:

```
> dengueseq[1:50]
```

[1] "a" "g" "t" "t" "g" "t" "t" "a" "g" "t" "c" "t" "a" "c" "g" "t" "g" "g" "a"

[20] "c" "c" "g" "a" "c" "a" "a" "g" "a" "a" "c" "a" "g" "t" "t" "t" "c" "g" "a"

[39] "a" "t" "c" "g" "g" "a" "a" "g" "c" "t" "t" "g"

Note that *dengueseq[1:50]* refers to the elements of the vector *dengueseq* with indices from 1-50. These elements contain the first 50 nucleotides of the DEN-1 Dengue virus sequence.

WRITING SEQUENCE DATA OUT AS A FASTA FILE

If you have retrieved a sequence from the NCBI database using the "*getncbiseq()*" function, you may want to save the sequence to a FASTA-format file on your computer, in case you need the sequence for further analyses (either in **R** or in other software).

You can write out a sequence to a FASTA-format file in **R** by using the "*write.fasta()*" function from the **SeqinR R** package. The *write.fasta()* function requires that you tell it the name of the output file using the "*file.out*" argument (input). You also need to specify the **R** variable that contains the sequence using the "*sequences*" argument, and the name that you want to give to the sequence using the "*names*" argument. For example, if you have stored the DEN-1 Dengue virus sequence in a vector *dengueseq*, to write out the sequence to a FASTA-format file called "den1.fasta" that contains the sequence labelled as "DEN-1", you can type:

> write.fasta(names="DEN-1", sequences=dengueseq, file.out="den1.fasta")

READING SEQUENCE DATA INTO R

Using the **SeqinR** package in **R**, you can easily read a DNA sequence from a FASTA file into **R**. For example, we described above how to retrieve the DEN-1 Dengue virus genome sequence from the NCBI database, or from **R** using the *getncbiseq()* function, and save it in a FASTA format file (e.g., "den1.fasta"). You can read this FASTA format file into **R** using the *read.fasta()* function from the **SeqinR R** package:

```
> library("seqinr")
> dengue <- read.fasta(file = "den1.fasta")
```

Note that **R** expects the files that you read in (e.g., "den1.fasta") to be in the "My Documents" folder on your computer, so if you stored "den1.fasta" somewhere else, you will have to move or copy it into "My Documents." The above command reads the contents of the fasta format file den1.fasta into an **R** object called *dengue*. The variable *dengue* is an **R** list object. As explained above, a *list* is an R object that is like a vector, but can contain elements that are numeric and/or contain characters. In this case, the list *dengue* contains information from the FASTA file that you have read in (i.e., the name given to the dengue sequence in the FASTA file, and the DNA sequence itself). In fact, the first element of the list object *dengue* contains the DNA sequence. We can access the elements of an **R** list object using double square brackets. Thus, we can store the DNA sequence for DEN-1 Dengue virus in a variable *dengueseq* by typing:

```
> dengueseq <- dengue[[1]]
```

Now the variable *dengueseq* is a vector containing the nucleotide sequence.

Length of a DNA sequence

Once you have retrieved a DNA sequence, we can obtain some simple statistics to describe that sequence, such as the sequence's total length in nucleotides. In the above example, we retrieved the DEN-1 Dengue virus genome sequence, and stored it in the vector variable *dengueseq*. To subsequently obtain the length of the genome sequence, we would use the *length()* function, typing:

```
> length(dengueseq)
[1] 10735
```

The *length()* function will give you back the length of the sequence stored in variable *dengueseq*, in nucleotides. The *length()* function actually gives the number of elements in the input vector that you pass to it, which in this case in the number of elements in the vector *dengueseq*. Since each element of the vector *dengueseq* contains one nucleotide of the DEN-1 Dengue virus sequence, the result for the DEN-1 Dengue virus genome tells us the length of its genome sequence (i.e., 10735 nucleotides long).

Base composition of a DNA sequence

An obvious first analysis of any DNA sequence is to count the number of occurrences of the four different nucleotides ("A", "C", "G", and "T") in the sequence. This can be done using the *table()* function. For example, to find the number of As, Cs, Gs, and Ts in the DEN-1 Dengue virus sequence (which you have put into vector variable *dengueseq*, using the commands above), you would type:

```
> table(dengueseq)
dengueseq
  a    c    g    t
3426 2240 2770 2299
```

This means that the DEN-1 Dengue virus genome sequence has 3426 As, 2240 Cs, 2770 Gs and 2299 Ts.

GC Content of DNA

One of the most fundamental properties of a genome sequence is its GC content, the fraction of the sequence that consists of Gs and Cs, i.e., the %(G+C). The GC content can be calculated as the percentage of the bases in the genome that are Gs or Cs. That is,

GC content = (number of Gs + number of Cs)*100/(genome length)

For example, if the genome is 100 bp, and 20 bases are Gs and 21 bases are Cs, then the GC content is (20 + 21)*100/100 = 41%.

You can easily calculate the GC content based on the number of As, Gs, Cs, and Ts in the genome sequence. For example, for the DEN-1 Dengue virus genome sequence, we know from using the table() function above that the genome contains 3426 As, 2240 Cs, 2770 Gs and 2299 Ts. Therefore, we can calculate the GC content using the command:

```
> (2240+2770)*100/(3426+2240+2770+2299)
[1] 46.66977
```

Alternatively, if you are feeling lazy, you can use the *GC()* function in the SeqinR package, which gives the fraction of bases in the sequence that are Gs or Cs.

```
> GC(dengueseq)
[1] 0.4666977
```

The result above means that the fraction of bases in the DEN-1 Dengue virus genome that are Gs or Cs is 0.4666977. To convert the fraction to a percentage, we have to multiply by 100, so the GC content as a percentage is 46.66977%.

DNA words

As well as the frequency of each of the individual nucleotides ("A", "G", "T", "C") in a DNA sequence, it is also interesting to know the frequency of longer DNA "words." The individual nucleotides are DNA words that are 1 nucleotide long, but we may also want to find out the frequency of DNA words that are 2 nucleotides long (i.e., "AA", "AG", "AC", "AT", "CA", "CG", "CC", "CT", "GA", "GG", "GC", "GT", "TA", "TG", "TC", and "TT"), 3 nucleotides long (e.g., "AAA", "AAT", "ACG", etc.), 4 nucleotides long, etc.

To find the number of occurrences of DNA words of a particular length, we can use the count() function from the R SeqinR package. For example, to find the number of occurrences of DNA words that are 1 nucleotide long in the sequence *dengueseq*, we type:

```
> count(dengueseq, 1)
  a c g t
 3426 2240 2770 2299
```

As expected, this gives us the number of occurrences of the individual nucleotides. To find the number of occurrences of DNA words that are 2 nucleotides long, we type:

```
> count(dengueseq, 2)
  aa ac ag at ca cc cg ct ga gc gg gt ta tc tg tt
  1108 720 890 708 901 523 261 555 976 500 787 507 440 497
832 529
```

Note that by default the count() function includes all overlapping DNA words in a sequence. Therefore, for example, the sequence "ATG" is considered to contain two words that are two nucleotides long: "AT" and "TG".

If you type help('count'), you will see that the result (output) of the function count() is a *table* object. This means that you can use double square brackets to extract the values of elements from the table. For example, to extract the value of the third element (the number of Gs in the DEN-1 Dengue virus sequence), you can type:

```
> denguetable <- count(dengueseq,1)
> denguetable[[3]]
 [1] 2770
```

The command above extracts the third element of the table produced by count(dengueseq,1), which we have stored in the table variable *denguetable*.

Alternatively, you can find the value of the element of the table in column "g" by typing:

```
> denguetable[["g"]]
 [1] 2770
```

SUMMARY

In the previous sections, you have learnt to use the following R functions:

- *length()* for finding the length of a vector or list
- *table()* for printing out a table of the number of occurrences of each type of item in a vector or list.

These functions belong to the standard installation of **R**.

You have also learnt the following **R** functions that belong to the **SeqinR** package:

- *GC()* for calculating the GC content for a DNA sequence
- *count()* for calculating the number of occurrences of DNA words of a particular length in a DNA sequence

REFERENCES

- *Little Book of R for Bioinformatics* by Avril Coghlan, Wellcome Trust Sanger Institute, Cambridge, U.K (**2016**).
- Charif D & Lobry JR. "SeqinR 1.0-2: a contributed package to the R project for Statistical Computing Devoted to Biological Sequences Retrieval and Analysis in Structural approaches to sequence evolution: Molecules, networks, populations (**2007**).

5

SEQUENCE STATISTICS

DNA SEQUENCE STATISTICS

We will delve a little deeper into sequence statistics. We will use DNA sequence to illustrate different kind of regular statistics calculations in **R**. This builds on the basics built in the previous chapter. Hence, the reader is requested to read the Chapter 4 before taking up the present Chapter. We start this chapter by introducing some handy functions for quick numerical computations.

Briefly, you learnt about variables in **R**, such as scalars, vectors, and lists in the previous chapter. You also learnt how to use functions to carry out operations on variables, for example, using the *log10()* function to calculate the *log* to the *base 10* of a scalar variable *x*, or using the *mean()* function to calculate the average of the values in a vector variable *myvector*:

```
> x <- 100
> log10(x)
[1] 2
> myvector <- c(30,16,303,99,11,111)
> mean(myvector)
```

```
[1] 95
```

You also learnt that you can extract an element of a vector by typing the vector name with the index of that element given in square brackets. For example, to get the value of the 3rd element in the vector *myvector*, we type:

```
> myvector[3]
[1] 303
```

A useful function in **R** is the *seq()* function, which can be used to create a sequence of numbers that run from a particular number to another particular number. For example, if we want to create the sequence of numbers from 1-100 in steps of 1 (i.e., 1, 2, 3, 4, ... 97, 98, 99, 100), we can type:

```
> seq(1, 100, by = 1)
  1  2  3  4  5  6  7  8  9 10 11 12 13 14 15 16 17 18
[19] 19 20 21 22 23 24 25 26 27 28 29 30 31 32 33 34 35 36
[37] 37 38 39 40 41 42 43 44 45 46 47 48 49 50 51 52 53 54
[55] 55 56 57 58 59 60 61 62 63 64 65 66 67 68 69 70 71 72
[73] 73 74 75 76 77 78 79 80 81 82 83 84 85 86 87 88 89 90
[91] 91 92 93 94 95 96 97 98 99 100
```

We can change the step size by altering the value of the "*by*" argument given to the function *seq()*. For example, if we want to create a sequence of numbers from 1-100 in steps of 2 (i.e., 1, 3, 5, 7, ... 97, 99), we can type:

```
> seq(1, 100, by = 2)
[1] 1 3 5 7 9 11 13 15 17 19 21 23 25 27 29 31 33 35 37 39 41 43 45 47 49
[26] 51 53 55 57 59 61 63 65 67 69 71 73 75 77 79 81 83 85 87 89 91 93 95 97 99
```

In **R**, just as in programming languages such as **Python**, it is possible to write a *for loop* to carry out the same command several times. For example, if we want to print out the square of each number between 1 and 10, we can write the following for loop:

```
> for (i in 1:10) { print (i*i) }
```

```
[1] 1
[1] 4
[1] 9
[1] 16
[1] 25
[1] 36
[1] 49
[1] 64
[1] 81
[1] 100
```

In the *for loop* above, the variable *i* is a counter for the number of cycles through the loop. In the first cycle through the loop, the value of *i* is 1, and so *i* * *i* = 1 is printed out. In the second cycle through the loop, the value of *i* is 2, and so *i* * *i* = 4 is printed out. In the third cycle through the loop, the value of *i* is 3, and so *i* * *i* = 9 is printed out. The loop continues until the value of *i* is 10. In the tenth cycle through the loop, the value of *i* is 10, and so *i* * *i* = 100 is printed out. Note that the commands that are to be carried out at each cycle of the *for loop* must be enclosed within curly brackets ("{" and "}").

You can also give a *for loop* a vector of numbers containing the values that you want the counter *i* to take in subsequent cycles. For example, you can make a vector *avector* containing the numbers 2, 9, 100, and 133, and write a *for loop* to print out the square of each number in vector *avector*:

```
> avector <- c(2, 9, 100, 133)
> for (i in avector) { print (i*i) }
[1] 4
[1] 81
[1] 10000
[1] 17689
```

How can we use a *for loop* to print out the square of every second number between 1 and 10? The answer is to use the *seq()* function to tell the *for loop* to take every second number between 1 and 10:

```
> for (i in seq(1, 10, by = 2)) { print (i*i) }
```

[1] 1

[1] 9

[1] 25

[1] 49

[1] 81

In the first cycle of this loop, the value of *i* is 1, and so *i* * *i* = 1 is printed out. In the second cycle through the loop, the value of *i* is 3, and so *i* * *i* = 9 is printed out. The loop continues until the value of *i* is 9. In the fifth cycle through the loop, the value of *i* is 9, and so *i* * *i* = 81 is printed out.

R allows the production of a variety of plots, including scatterplots, histograms, piecharts, and boxplots. For example, if you have two vectors of numbers *myvector1* and *myvector2*, you can plot a scatterplot of the values in *myvector1* against the values in *myvector2* using the *plot()* function. If you want to label the axes on the plot, you can do this by giving the *plot()* function values for its optional arguments *xlab* and *ylab*:

```
> myvector1 <- c(10, 15, 22, 35, 43)
> myvector2 <- c(3, 3.2, 3.9, 4.1, 5.2)
> plot(myvector1, myvector2, xlab="myvector1", ylab="myvector2")
```

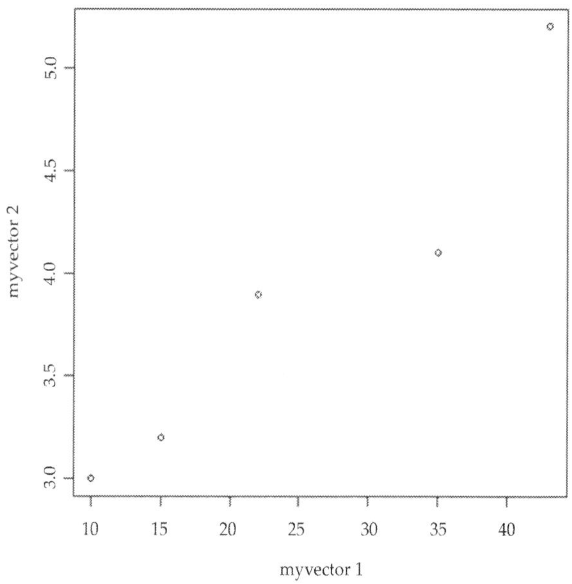

If you look at the help page for the *plot()* function, you will see that there are lots of optional arguments (inputs) that it can take that. For example, one optional argument is the *type* argument, that determines the type of the plot. By default, *plot()* will draw a dot at each data point, but if we set *type* to be "b", then it will also draw a line between each subsequent data point:

> plot(myvector1, myvector2, xlab="myvector1", ylab="myvector2", type="b")

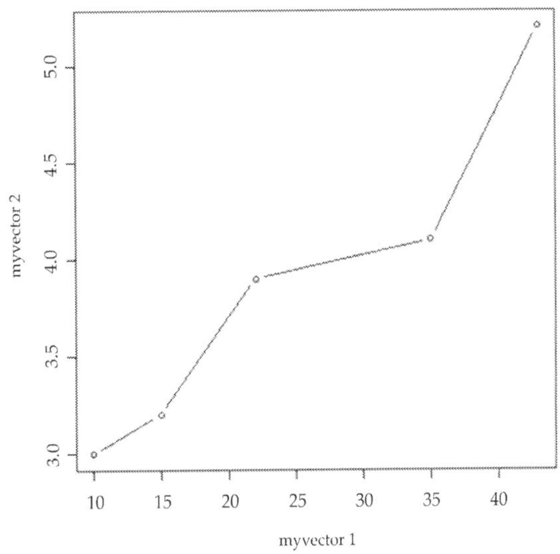

We have been using built-in **R** functions such as *mean ()*, *length ()*, *print()*, *plot()* etc. We can also create our own functions in **R** to do calculations that you want to carry out very often on different input data sets. For example, we can create a function to calculate the value of 20 plus the square of some input number:

> myfunction <- function(x) { return(20 + (x*x)) }

This function will calculate the square of a number (x), and then add 20 to that value. The *return ()* statement returns the calculated value. Once you have typed in this function, the function is then available for use. For example, we can use the function for different input numbers (e.g., 10, 25):

> myfunction(10)

[1] 120

```
> myfunction(25)
[1] 645
```

You can view the code that makes up a function by typing its name (without any parentheses). For example, we can try this by typing "*myfunction*":

```
> myfunction
function(x) { return(20 + (x*x)) }
```

When you are typing **R**, if you want to, you can write comments by writing the comment text after the "#" sign. This can be useful if you want to write some **R** commands that other people need to read and understand. **R** will ignore the comments when it is executing the commands. For example, you may want to write a comment to explain what the function *log10 ()* does:

```
> x <- 100
> log10(x) # Finds the log to the base 10 of variable x.
[1] 2
```

READING SEQUENCE DATA WITH SEQINR

In the previous chapter you learnt how to use to search for and download the sequence data for a given NCBI accession from the NCBI Sequence Database, either *via* the NCBI website or using the *getncbiseq()* function in **R**. For example, you could have downloaded the sequence data for a the DEN-1 Dengue virus sequence (NCBI accession NC_001477), and stored it on a file on your computer (e.g., "den1.fasta").

Once you have downloaded the sequence data for a particular NCBI accession, and stored it on a file on your computer, you can then read it into **R** by using *read.fasta* function from the **SeqinR R** package. For example, if you have stored the DEN-1 Dengue virus sequence in a file "den1.fasta", you can type:

```
> library("seqinr")                              # Load the SeqinR package.
> dengue <- read.fasta(file = "den1.fasta")   # Read in the file "den1.
fasta".
> dengueseq <- dengue[[1]]                        # Put the sequence in a vector
called "dengueseq".
```

Once you have retrieved a sequence from the NCBI Sequence Database and stored it in a vector variable such as *dengueseq* in the example above, it is possible to extract subsequence of the sequence by type the name of the vector (e.g., *dengueseq*) followed by the square brackets containing the indices for those nucleotides. For example, to obtain nucleotides 452-535 of the DEN-1 Dengue virus genome, we can type:

```
> dengueseq[452:535]
 [1] "c" "g" "a" "g" "g" "g" "g" "g" "a" "g" "a" "g" "c" "c" "g" "c" "a" "c" "a"
[20] "t" "g" "a" "t" "a" "g" "t" "t" "a" "g" "c" "a" "a" "g" "c" "a" "g" "g" "a"
[39] "a" "a" "g" "a" "g" "g" "a" "a" "a" "a" "t" "c" "a" "c" "t" "t" "t" "t" "g"
[58] "t" "t" "t" "a" "a" "g" "a" "c" "c" "t" "c" "t" "g" "c" "a" "g" "g" "t" "g"
[77] "t" "c" "a" "a" "c" "a" "t" "g"
```

LOCAL VARIATION IN GC CONTENT

You know now how to calculate the GC content of a genome sequence (percentage of nucleotides in a genome sequence that are Gs or Cs), you can use the *GC()* function in the **SeqinR** package. For example, to find the GC content of the DEN-1 Dengue virus sequence that we have stored in the vector *dengueseq*, we can type:

```
> GC(dengueseq)
[1] 0.4666977
```

The output of the *GC ()* is the fraction of nucleotides in a sequence that are Gs or Cs, so to convert it to a percentage we need to multiply by 100. Thus, the GC content of the DEN-1 Dengue virus genome is about 0.467 or 46.7%.

Although the GC content of the whole DEN-1 Dengue virus genome sequence is about 46.7%, there is probably local variation in GC content within the genome. That is, some regions of the genome sequence may have GC contents quite a bit higher than 46.7%, while some regions of the genome sequence may have GC contents that are quite a big lower than 46.7%. Local fluctuations in GC content within the genome sequence can provide different interesting information, for example, they may reveal cases of horizontal transfer or reveal biases in mutation.

If a chunk of DNA has moved by horizontal transfer from the genome of a species with low GC content to a species with high GC content, the chunk of horizontally transferred DNA could be detected as a region of unusually low GC content in the high-GC recipient genome.

On the other hand, a region unusually low GC content in an otherwise high-GC content genome could also arise due to biases in mutation in that region of the genome, for example, if mutations from Gs/Cs to Ts/As are more common for some reason in that region of the genome than in the rest of the genome.

A SLIDING WINDOW ANALYSIS OF GC CONTENT

In order to study local variation in GC content within a genome sequence, we could calculate the GC content for small chunks of the genome sequence. The DEN-1 Dengue virus genome sequence is 10735 nucleotides long. To study variation in GC content within the genome sequence, we could calculate the GC content of chunks of the DEN-1 Dengue virus genome, for example, for each 2000-nucleotide chunk of the genome sequence:

> GC(dengueseq[1:2000]) # Calculate the GC content of nucleotides 1-2000 of the Dengue genome

[1] 0.465

> GC(dengueseq[2001:4000]) # Calculate the GC content of nucleotides 2001-4000 of the Dengue genome

[1] 0.4525

> GC(dengueseq[4001:6000]) # Calculate the GC content of nucleotides 4001-6000 of the Dengue genome

[1] 0.4705

> GC(dengueseq[6001:8000]) # Calculate the GC content of nucleotides

6001-8000 of the Dengue genome

```
[1] 0.479
> GC(dengueseq[8001:10000]) # Calculate the GC content of nucleotides 8001-10000 of the Dengue genome
[1] 0.4545
> GC(dengueseq[10001:10735]) # Calculate the GC content of nucleotides 10001-10735 of the Dengue genome
[1] 0.4993197
```

From the output of the above calculations, we see that the region of the DEN-1 Dengue virus genome from nucleotides 1-2000 has a GC content of 46.5%, while the region of the Dengue genome from nucleotides 2001-4000 has a GC content of about 45.3%. Thus, there seems to be some local variation in GC content within the Dengue genome sequence.

Instead of typing in the commands above to tell R to calculate the GC content for each 2000-nucleotide chunk of the DEN-1 Dengue genome, we can use a *for loop* to carry out the same calculations, but by typing far fewer commands. That is, we can use a *for loop* to take each 2000-nucleotide chunk of the DEN-1 Dengue virus genome, and to calculate the GC content of each 2000-nucleotide chunk. Below we will explain the following *for loop* that has been written for this purpose:

```
> starts <- seq(1, length(dengueseq)-2000, by = 2000)
> starts
[1]    1 2001 4001 6001 8001
> n <- length(starts)    # Find the length of the vector "starts"
> for (i in 1:n) {
      chunk <- dengueseq[starts[i]:(starts[i]+1999)]
      chunkGC <- GC(chunk)
      print (chunkGC)
   }
[1] 0.465
[1] 0.4525
[1] 0.4705
[1] 0.479
```

[1] 0.4545

The command “starts <- seq(1, length(dengueseq)-2000, by = 2000)” stores the result of the *seq()* command in the vector *starts*, which contains the values 1, 2001, 4001, 6001, and 8001.

We set the variable *n* to be equal to the number of elements in the vector *starts*, so it will be 5 here, since the vector *starts* contains the five elements 1, 2001, 4001, 6001 and 8001. The line “*for (i in 1:n)*” means that the counter *i* will take values of 1-5 in subsequent cycles of the *for loop*. The *for loop* above is spread over several lines. However, **R** will not execute the commands within the *for loop* until you have typed the final “}” at the end of the *for loop* and pressed “Return”.

Each of the three commands within the *for loop* are carried out in each cycle of the loop. In the first cycle of the loop, *i* is 1, the vector variable *chunk* is used to store the region from nucleotides 1-2000 of the Dengue virus sequence, the GC content of that region is calculated and stored in the variable *chunkGC*, and the value of *chunkGC* is printed out.

In the second cycle of the loop, *i* is 2, the vector variable *chunk* is used to store the region from nucleotides 2001-4000 of the Dengue virus sequence, the GC content of that region is calculated and stored in the variable *chunkGC*, and the value of *chunkGC* is printed out. The loop continues until the value of *i* is 5. In the fifth cycle through the loop, the value of *i* is 5, and so the GC content of the region from nucleotides 8001-10000 is printed out.

Note that we stop the loop when we are looking at the region from nucleotides 8001-10000, instead of continuing to another cycle of the loop where the region under examination would be from nucleotides 10001-12000. The reason for this is because the length of the Dengue virus genome sequence is just 10735 nucleotides, so there is not a full 2000-nucleotide region from nucleotide 10001 to the end of the sequence at nucleotide 10735.

The above analysis of local variation in GC content is what is known as a *sliding window analysis of GC content*. By calculating the GC content in each 2000-nucleotide chunk of the Dengue virus genome, you are effectively sliding a 2000-nucleotide *window* along the DNA sequence from start to end, and calculating the GC content in each non-overlapping window (chunk of DNA).

Note that this sliding window analysis of GC content is a slightly simplified version of the method usually carried out by bioinformaticians. In this simplified version, we have calculated the GC content in non-overlapping windows along a DNA sequence. However, it is more usual to calculate GC content in overlapping windows along a sequence, although that makes the code slightly more complicated.

A SLIDING WINDOW PLOT OF GC CONTENT

It is common to use the data generated from a sliding window analysis to create a *sliding window plot of GC content*. To create a sliding window plot of GC content, you plot the local GC content in each window of the genome, versus the nucleotide position of the start of each window. We can create a sliding window plot of GC content by typing:

```
> starts <- seq(1, length(dengueseq)-2000, by = 2000)
> n <- length(starts)    # Find the length of the vector "starts"
> chunkGCs <- numeric(n) # Make a vector of the same length as vector "starts", but just containing zeroes
> for (i in 1:n) {
      chunk <- dengueseq[starts[i]:(starts[i]+1999)]
      chunkGC <- GC(chunk)
      print(chunkGC)
      chunkGCs[i] <- chunkGC
  }
> plot(starts,chunkGCs,type="b",xlab="Nucleotide start position",ylab="GC content")
```

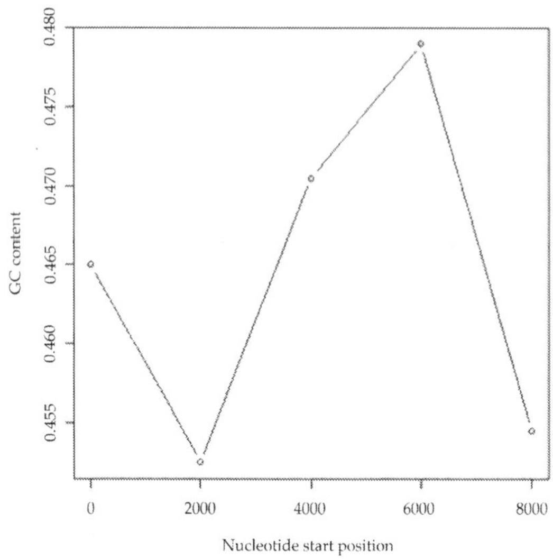

In the code above, the line "chunkGCs <- numeric (n)" makes a new vector *chunkGCs* which has the same number of elements as the vector *starts* (5 elements here). This vector *chunkGCs* is then used within the *for loop* for storing the GC content of each chunk of DNA.

After the loop, the vector *starts* can be plotted against the vector *chunkGCs* using the *plot ()* function, to get a plot of GC content against nucleotide position in the genome sequence. This is a *sliding window plot of GC content.*

You may want to use the code above to create sliding window plots of GC content of different species' genomes, using different window sizes. Therefore, it makes sense to write a function to do the sliding window plot, that can take the window size that the user wants to use and the sequence that the user wants to study as arguments (inputs):

```
> slidingwindowplot <- function(windowsize, inputseq)
{
  starts <- seq(1, length(inputseq)-windowsize, by = windowsize)
  n <- length(starts)    # Find the length of the vector "starts"
  chunkGCs <- numeric(n) # Make a vector of the same length as vector
"starts", but just containing zeroes
  for (i in 1:n) {
```

```
    chunk <- inputseq[starts[i]:(starts[i]+windowsize-1)]
    chunkGC <- GC(chunk)
    print(chunkGC)
    chunkGCs[i] <- chunkGC
  }
    plot(starts,chunkGCs,type="b",xlab="Nucleotide start position",ylab="GC content")
}
```

This function will make a sliding window plot of GC content for a particular input sequence *inputseq* specified by the user, using a particular window size *windowsize* specified by the user. Once you have typed in this function, you can use it again and again to make sliding window plots of GC contents for different input DNA sequences, with different window sizes. For example, you could create two different sliding window plots of the DEN-1 Dengue virus genome sequence, using window sizes of 3000 and 300 nucleotides, respectively:

```
> slidingwindowplot(3000, dengueseq)
```

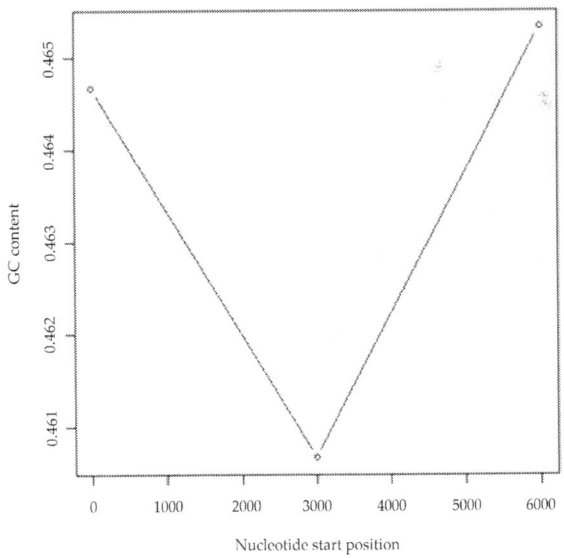


```
> slidingwindowplot(300, dengueseq)
```

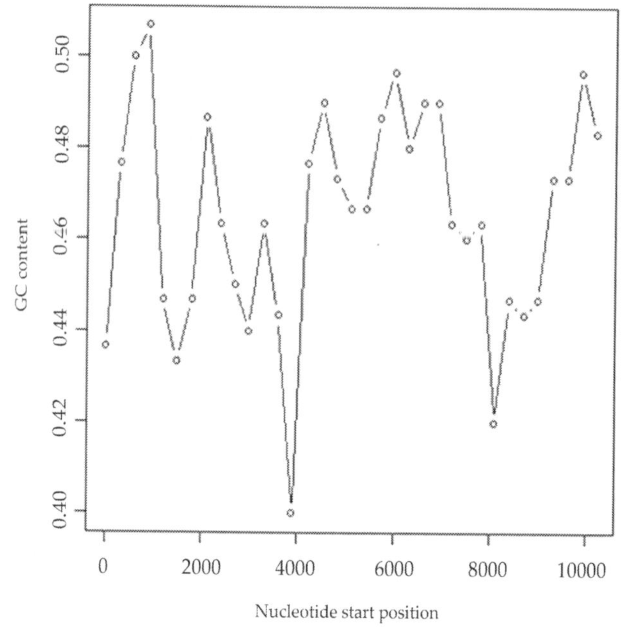

OVER-REPRESENTED AND UNDER-REPRESENTED DNA WORDS

In the previous sections, you learnt that the *count ()* function in the **SeqinR R** package can calculate the frequency of all DNA words of a certain length in a DNA sequence. For example, if you want to know the frequency of all DNA words that are 2 nucleotides long in the Dengue virus genome sequence, you can type:

```
> count(dengueseq, 2)
aa ac ag at ca cc cg ct ga gc gg gt ta tc tg tt
1108 720 890 708 901 523 261 555 976 500 787 507 440 497
832 529
```

It is interesting to identify DNA words that are two nucleotides long ("dinucleotides", i.e., "AT", "AC", etc.) that are over-represented or under-represented in a DNA sequence. If a particular DNA word is *over-represented* in a sequence, it means that it occurs many more times in the sequence than you would have expected by chance. Similarly, if a particular DNA word is *under-represented* in a sequence, it means it occurs far fewer times in the sequence than you would have expected.

A statistic called ρ (Rho) is used to measure how *over-* or *under-represented* a particular DNA word is in a sequence. For a 2-nucleotide (dinucleotide) DNA word ρ is calculated as:

$$\rho(xy) = f_{xy}/(f_x * f_y),$$

where f_{xy} and f_x are the frequencies of the DNA words xy and x in the DNA sequence under study. For example, the value of for the DNA word "TA" can be calculated as:

$$\rho(TA) = f_{TA}/(f_T * f_A),$$

where f_{TA}, f_T and f_A are the frequencies of the DNA words "TA", "T" and "A" in the DNA sequence.

The idea behind the ρ statistic is that, if a DNA sequence had a frequency f_x of a 1-nucleotide DNA word x, and a frequency f_y of a 1-nucleotide DNA word y, then we expect the frequency of the 2-nucleotide DNA word xy to be $f_x * f_y$. That is, the frequencies of the 2-nucleotide DNA words in a sequence are expected to be equal the products of the specific frequencies of the two nucleotides that compose them. If this were true, then ρ would be equal to 1. If we find that ρ is much greater than 1 for a particular 2-nucleotide word in a sequence, it indicates that that 2-nucleotide word is much more common in that sequence than expected (i.e., it is *over-represented*).

For example, say that your input sequence has only 5% Ts (i.e., $f_T = 0.05$). In a random DNA sequence with 5% Ts, you would expect to see the word "TT" very infrequently. In fact, we would only expect 0.05 * 0.05=0.0025 (0.25%) of 2-nucleotide words to be TTs (i.e., we expect $f_{TT} = f_T * f_T$). This is because Ts are rare, so they are expected to be adjacent to each other very infrequently if the few Ts are randomly scattered throughout the DNA. Therefore, if you see lots of TT 2-nucleotide words in your real input sequence (e.g., $f_{TT} = 0.3$, so $\rho = 0.3/0.0025 = 120$), you would suspect that natural selection has acted to increase the number of occurrences of the TT word in the sequence (presumably because it has some beneficial biological function).

To find over-represented and under-represented DNA words that are 2 nucleotides long in the DEN-1 Dengue virus sequence, we can calculate the ρ statistic for each 2-nucleotide word in the sequence. For example, given the number of occurrences of the individual nucleotides A, C, G and T in the Dengue sequence, and the number of occurrences of the DNA word GC in the sequence (500, from above), we can calculate the value of ρ for the 2-nucleotide DNA word "GC", using the formula

$$\rho(GC) = f_{GC}/(f_G * f_C),$$

where f_{GC}, f_G and f_C are the frequencies of the DNA words "GC", "G" and "C" in the DNA sequence:

```
> count(dengueseq, 1) # Get the number of occurrences of 1-nucleotide DNA words
   a    c    g    t
3426 2240 2770 2299
> 2770/(3426+2240+2770+2299) # Get fG
[1] 0.2580345
> 2240/(3426+2240+2770+2299) # Get fC
[1] 0.2086633
> count(dengueseq, 2) # Get the number of occurrences of 2-nucleotide DNA words
  aa  ac  ag  at  ca  cc  cg  ct  ga  gc  gg  gt  ta  tc  tg  tt
1108 720 890 708 901 523 261 555 976 500 787 507 440 497 832 529
> 500/(1108+720+890+708+901+523+261+555+976+500+787+507+440+497+832+529) # Get fGC
[1] 0.04658096
> 0.04658096/(0.2580345*0.2086633) # Get rho(GC)
[1] 0.8651364
```

We calculate a value of ρ(GC) of approximately 0.865. This means that the DNA word "GC" is about 0.865 times as common in the DEN-1 Dengue virus sequence than expected. That is, it seems to be slightly under-represented.

Note that if the ratio of the observed to expected frequency of a particular DNA word is very low or very high, then we would suspect that there is a statistical under-representation or over-representation of that

DNA word. However, to be sure that this *over*- or *under*-representation is statistically significant, we would need to do a *statistical test*. We will not deal with the topic of how to carry out the statistical test here.

SUMMARY

In this chapter, you will have learnt to use the following R functions:

- *seq()* for creating a sequence of numbers
- *print()* for printing out the value of a variable
- *plot()* for making a plot (e.g., a scatterplot)
- *numeric()* for making a numeric vector of a particular length
- *function()* for making a function

All of these functions belong to the standard installation of **R**. You also learnt how to use *for loops* to carry out the same operation again and again, each time on different inputs.

REFERENCES

1. *Little Book of R for Bioinformatics* by Avril Coghlan, Wellcome Trust Sanger Institute, Cambridge, U.K (2016).
2. *Applied Statistics for Bioinformatics using R* by Wim P. Krinjen (**2009**).
3. *Introduction to Computational Genomics* by Cristianini N & Hahn MW, (Cambridge University Press, **2006**).

6

MULTIPLE SEQUENCE ALIGNMENTS

We learned about *pairwise*, *global* and *local* sequence alignments in the previous chapter. We also looked at the tools which can be used to accomplish these tasks. In practice, a scientist will be comparing more than two sequences. A BLAST search can yield more than one hit. How do you understand these these results? Does the results tell you anything about its evolution and relatedness with each other? Instead of performing pairwise alignments, it would save time to perform sequence alignments of multiple sequences and compare them all at once. *Multiple sequence alignment* (MSA) is frequently used to study groups of related genes or proteins and infer evolutionary relationships. We will learn how to perform MSA and discuss some applications like *phylogenetic trees*.

We start by discussing evolution and get familiar with the terms and definitions in the subject area.

EVOLUTION - INTRODUCTION

The theory of evolution accepted by most biologists of present scientific community is that proposed by **Charles Darwin**. Darwin proposed that variability is a feature of every generation of living creatures. The individuals whose variations predispose them to survive in their environment are the ones who reproduce most successfully and pass on their traits in greater numbers. In layman's terms this is known as '*survival of the fittest*'. In the light of this theory, it has been hypothesized that the diversity of life forms on Earth is due to variability. This divergence of physical and genetic traits from one common ancestral unicellular organism to subsequent generations creates the observable variability in different biological niches.

Evolution was initially proposed to explain the physical traits of living organisms. This concept was extended to discuss evolution at the level of DNA and protein sequences. Although the replication of DNA sequence is a very accurate process, small replication errors accumulate over time, along with radiation damage and other mutations or alterations of the genomic sequence. This is known in a layman's language as *mutations*. Instead of evolutionary pressure selecting organisms based on morphological traits, selection occurs at the level of mutations. Consequently, the only mutations observed in genes from healthy organisms are those that don't result in the organisms' death.

The base composition of DNA is prone to chemical modifications through above processes. The chemical processes of molecular evolution are responsible for more than just giving rise to species differences. Evolutionary change can occur within the genome of a single species as well. Hence, two different situations arise when discussing molecular evolution.

- *Orthologs* are genes that are evolutionarily related, share a function, and have diverged by speciation.
- *Paralogs*, on the other hand, have a common ancestor but have diverged by gene duplication and no longer have a common functional role.

In other words, *orthologs* have the same function but occur in different species, while *paralogs* exist in the **same genome** but have different functions. A sequence database search may return both *orthologs* and *paralogs*. Depending on the objectives of your study, you probably will not want to treat them all as members of the same set.

CLUSTALW AND MULTIPLE SEQUENCE ALIGNMENT

One commonly used program for progressive multiple sequence alignment is **ClustalW**. The heuristic used in ClustalW is based on phylogenetic analysis. It can be understood to comprise four steps. These steps are described in simple terms below.

- A *pairwise* distance matrix for all the sequences to be aligned is generated. A guide tree is created using the neighbour-joining algorithm.
- Each of the most closely related pairs of sequences - the *outermost branches* of the tree - are aligned to each other using dynamic programming.
- Each new alignment is analyzed to *build* a *sequence profile.*
- Finally, *alignment profiles are aligned* to each other or to other sequences (depending on the topology of the tree) until a full alignment is built.

This strategy produces reasonable alignments under a range of conditions. It's not fool proof; for distantly related sequences, it can build on the inaccuracies of pairwise alignment and phylogenetic analysis. But for sequence sets with some recognizably related pairs, it builds on the strengths of these methods. Pairwise sequence alignment by dynamic programming is very accurate for closely related sequences regardless of which scoring matrix or penalty values are used. Phylogenetic analysis is relatively unambiguous for closely related sequences. Using multiple sequences to create profiles increases the accuracy of pairwise alignment for more distantly related sequences. There are many parameters involved in multiple sequence alignment. There are, of course, scoring matrices and gap penalties associated with the pairwise alignment steps. In addition, there are weighting parameters that alter the scoring matrix used in *sequence-profile* and *profile-profile* alignments. In **ClustalW**, these are set from the Multiple Alignment submenu or the Profile Structure Alignments submenu. In ClustalX, they are set from the Alignment pulldown menu. The pairwise alignment parameters are familiar and have the same meaning in multiple alignment as they do in pairwise alignment. The multiple alignment parameters include gap opening and gap extension penalties for the multiple alignment process—to be used when fine-tuning alignments—and a maximum allowable delay, in terms of sequence length, for the start of divergent sequences at the beginning of the alignment.

One of **ClustalW**'s heuristics is that, in protein sequence alignment, different scoring matrices are used for each alignment based on expected evolutionary distance. If two sequences are close neighbors in the tree, a scoring matrix optimized for close relationships aligns them. Distant neighbors are aligned using matrices optimized for distant relationships. Thus, when prompted to choose a series of matrices in the Multiple Alignment Parameters menu, it means just that: use BLOSUM62 for close relationships and BLOSUM45 for more distant relationships, rather than the same scoring matrix for all pairwise alignments. Another heuristic that **ClustalW** uses is *scalable gap penalties* for protein profile alignments. A gap opening next to a conserved hydrophobic residue can be penalized more heavily than a gap opening next to a hydrophilic residue. A gap opening too close to another gap can be penalized more heavily than an isolated gap. These parameters are set in the Protein Gap Parameters menu.

While **ClustalW** is simple to install and use on a Linux workstation, **ClustalX**, the X Windows-based graphical user interface for **ClustalW**, isn't so easy to compile. However, **ClustalX** runs in its own X window, has pull down menus, and allows viewing and plotting of multiple sequence alignments in a color-coded format. It also allows you to append sequences to an alignment one at a time, and to produce color PostScript output of specified sequence ranges in an alignment from different files, if desired, along with other convenient features. To install **ClustalX** on a Linux machine, you need:

- The **ClustalX** binaries
- The NCBI software toolkit source distribution
- The LessTif libraries

JALVIEW FOR VIEWING AND EDITING ALIGNMENTS

Jalview is written by Michele Clamp and available from EBI at:

www.ebi.ac.uk/~michele/jalview/contents.html

One can get installation instructions from the above website. The program can read an alignment file (.aln) from Clustal, among many other formats. One uses Jalview mostly as an alignment editor, although it has many other features.

SEQUENCE LOGOS

This is another way of viewing sequence alignments, developed by Tom Schneider at National Cancer Institute. This is more suited for short sequence regions, as in protein motifs. We will not be considering this in the present book.

There are many programs available, academic and commercial, for viewing and editing alignments. It is left to the reader to choose an editor that best suits his or her preferences.

PHYLOGENETIC ANALYSIS

This is quite a general and common problem in biology, mainly used to understand evolutionary relationships among proteins. Essentially, you will be addressing the question as to how species relate to one another and common ancestors. This is done by constructing trees, whose leaves represent present-day species and whose interior nodes represent hypothetical ancestors. These are called *phylogenetic trees*. Usually, we represent these relationships as a *dendrogram*. We will look at a simple example MSA and Phylogenetic Tree construction using **R**.

RETRIEVING A LIST OF SEQUENCES FROM UNIPROT

In previous chapters, you learnt how to search for DNA or protein sequences in sequence databases such as the NCBI database and UniProt, using the **SeqinR** package. Often times, it is useful to retrieve several sequences from UniProt at once if you have a list of UniProt accessions. The **R** function "*retrieveseqs()*" below is useful for this purpose:

```
> retrieveseqs <- function(seqnames,acnucdb)
  {
    myseqs <- list()  # Make a list to store the sequences
    require("seqinr")  # This function requires the SeqinR R package
    choosebank(acnucdb)
    for (i in 1:length(seqnames))
    {
      seqname <- seqnames[i]
      print(paste("Retrieving sequence",seqname,"..."))
      queryname <- "query2"
      query <- paste("AC=",seqname,sep="")
      query(`queryname`,`query`)
          seq <- getSequence(query2$req[[1]]) # Makes a vector "seq"
containing the sequence
      myseqs[[i]] <- seq
    }
    closebank()
    return(myseqs)
  }
```

You need to cut and paste this function into **R** to use it. As its input, you need to give the function a vector containing the accessions for the sequences you wish to retrieve, as well as the name of the ACNUC sub-database that the sequences should be retrieved from. In this case, we want to retrieve sequences from UniProt, so the sequences should be in the "swissprot" ACNUC sub-database.

The *retrieveseqs()* function returns a list variable, in which each element is a vector containing one of the sequences.

For example, to retrieve the protein sequences for UniProt accessions P06747, P0C569, O56773 and Q5VKP1 (the accessions for rabies virus phosphoprotein, Mokola virus phosphoprotein, Lagos bat virus phosphoprotein and Western Caucasian bat virus phosphoprotein, respectively), you can type:

> seqnames <- c("P06747", "P0C569", "O56773", "Q5VKP1") # Make a vector containing the names of the sequences

> seqs <- retrieveseqs(seqnames,"swissprot") # Retrieve the sequences and store them in list variable "seqs"

> length(seqs) # Print out the number of sequences retrieved

[1] 4

> seq1 <- seqs[[1]] # Get the first sequence

> seq1[1:20] # Print out the first 20 letters of the first sequence

[1] "M" "S" "K" "I" "F" "V" "N" "P" "S" "A" "I" "R" "A" "G" "L" "A" "D" "L" "E"

[20] "M"

> seq2 <- seqs[[2]] # Get the second sequence

> seq2[1:20] # Print out the first 20 letters of the second sequence

[1] "M" "S" "K" "D" "L" "V" "H" "P" "S" "L" "I" "R" "A" "G" "I" "V" "E" "L" "E"

[20] "M"

The commands above use the function *retrieveseqs()* to retrieve two UniProt sequences. The sequences are returned in a list variable *seqs*. To access the elements in an **R** list variable, you need to use double square brackets. Therefore, the second element of the list variable is accessed by typing *seqs[[2]]*. Each element of the list variable *seqs* contains a vector which stores one of the sequences.

Rabies virus is the virus responsible for rabies, which is classified by the WHO as a neglected tropical disease. Mokola virus and rabies virus are closely related viruses that both belong to a group of viruses called *Lyssaviruses*. Mokola virus causes a rabies-like infection in mammals including humans.

Once you have retrieved the sequences using *retrieveseqs()*, you can then use the function *write.fasta()* from the **SeqinR** package to write the sequences to a FASTA-format file. As its arguments (inputs), the *write.fasta()* function takes the list variable containing the sequences, and a vector containing the names of the sequences, and the name that you want to give to the FASTA-format file. For example:

```
> write.fasta(seqs, seqnames, file="phosphoproteins.fasta")
```

The command above will write the sequences in list variable *seqs* to a FASTA-format file called "phosphoproteins.fasta" in the "My Documents" folder on your computer.

INSTALLING THE CLUSTAL MULTIPLE ALIGNMENT SOFTWARE

Let us also take a look at the most commonly used tools in bioinformatics. A common task in bioinformatics is to download a set of related sequences from a database, and then to align those sequences using multiple alignment software. This is the first step in most phylogenetic analyses.

One commonly used multiple alignment software package is CLUSTAL. In order to build an alignment using CLUSTAL, you first need to install the CLUSTAL program on your computer.

To install CLUSTAL on your computer, you need to follow these steps:

- Go to the *http://www.clustal.org/download/current/* website.
- Right-click on the link to file **clustalx-Z.Z.Z-win.msi** (where Z represents some number) and choose "Save link as..." and then save the file in your "My Documents" folder.
- Once the file has downloaded, double-click on the icon for file **clustalx-Z.Z.Z-win.msi** (where Z is some number).
- You will be asked "Are you sure you want to run this software?" Press "Run".
- You will then see "Welcome to the ClustalX2 setup wizard." Press "Next".
- You will be asked where to install ClustalX2. Select your "My Documents" folder.
- Keep pressing 'yes' or 'Next' until the screen says "Completing the ClustalX2 setup wizard." Then press "Finish".

CLUSTAL should now be installed on your computer.

CREATING A MULTIPLE ALIGNMENT OF PROTEIN, DNA OR MRNA SEQUENCES USING CLUSTAL

Once you have installed CLUSTAL, you can now align your sequences using CLUSTAL by following these steps:

- Go to the "Start" menu on the bottom left of your Windows screen. Select "All Programs" from the menu, then select "ClustalX2" from the menu that appears. This will start up CLUSTAL.
- The CLUSTAL window should appear. To load the DNA or protein sequences that you want to align into CLUSTAL, go to the CLUSTAL "File" menu, and choose "Load sequences".
- Select the FASTA-format file containing your sequences (e.g., phosphoproteins. fasta) to load it into CLUSTAL.
- This should read the sequences into CLUSTAL. They have not been aligned yet, but will be displayed in the CLUSTAL window.
- You can use the scrollbar on the right to scroll down and look at all the sequences. You can use the scrollbar on the bottom to scroll from left to right, and look along the length of the sequences.
- Before you align the sequences using CLUSTAL, you need to tell CLUSTAL to make the output alignment file in PHYLIP alignment format, so that you can read it into R. To do this, go to the "Alignment" menu in CLUSTAL, choose "Output Format Options." A form will appear, and in this form you should select "PHYLIP format" and deselect "CLUSTAL format", and then press "OK".
- To now align the sequences using CLUSTAL, go to the CLUSTAL "Alignment" menu, and choose "Do Complete Alignment".
- A menu box will pop up, asking you where to save the output guide-tree file (e.g., "phosphoproteins.dnd") and the output alignment file (called "phosphoproteins.phy"). You should

choose to save them in your "My Documents" folder (so that you can easily read them into R from "My Documents" at a later stage).

- CLUSTAL will now align the sequences. This will take a couple of minutes (e.g., 2-5 minutes). You will see that at the bottom of the CLUSTAL window, it tells you which pair of sequences it is aligning at a particular point in time. If the numbers keep changing, it means that CLUSTAL is still working away, and the alignment is not finished yet. Be patient!

Once CLUSTAL has finished making the alignment, it will be displayed in the CLUSTAL window. For example, here is the CLUSTAL alignment for rabies virus phosphoprotein, Mokola virus phosphoprotein, and Lagos bat virus phosphoprotein:

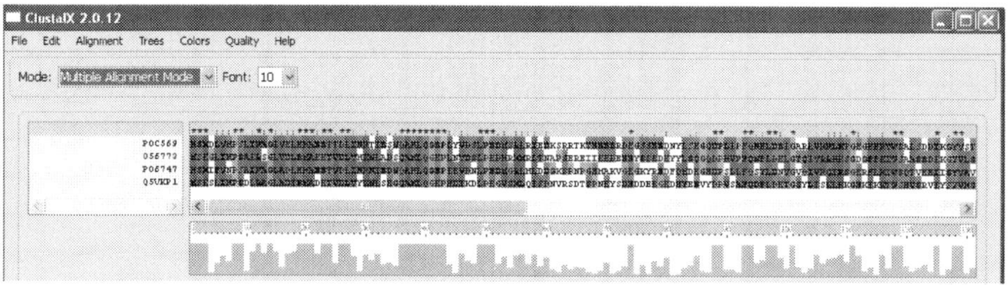

The alignment displayed in CLUSTAL has a row for each of your sequences. CLUSTAL colours sets of chemically similar amino acids in similar colours. For example, tyrosine (Y) is coloured blue-green, while the chemically similar amino acid phenylalanine (F) is coloured blue. You can scroll to the right and left along the alignment using the scrollbar at the bottom of the Jalview window.

Below the alignment, you can see a grey plot, showing the level of conservation at each point of the sequence. This shows a high grey bar if the conservation in a region is high (there is high percent identity between the sequence), and a low grey bar if it is low (there is low percent identity). This can give you an idea of which are the best conserved regions of the alginment.

For example, for the alignment of the four virus phosphoproteins, we can see that the region in alignment columns 35 to 45 approximately is very well conserved, while the region in alignment columns 60 to 70 is poorly conserved.

The CLUSTAL alignment will have been saved in a file in your "My Documents" folder called "something.phy" (e.g., phosphoproteins.phy). This is a PHYLIP-format alignment file, which you can now read into R for further analysis.

READING A MULTIPLE ALIGNMENT FILE INTO R

To read a sequence alignment into **R** from a file, you can use the *read.alignment()* function in the **SeqinR** package. For example, to read in the multiple sequence alignment of the virus phosphoproteins into **R**, we type:

> virusaln <- read.alignment(file = "phosphoproteins.phy", format = "phylip")

The *virusaln* variable is a list variable that stores the alignment.

An **R** list variable can have named elements, and you can access the named elements of a list variable by typing the variable name, followed by "$", followed by the name of the named element.

The list variable *virusaln* has named elements "nb", "nam", "seq", and "com".

In fact, the named element "seq" contains the alignment, which you can view by typing:

> virusaln$seq

[[1]]

[1] "mskdlvhpsliragivelemaeettdlinrtiesnqahlqgeplyvdslped msrlriedksrrtk...

[[2]]

[1] "mskglihpsairsglvdlemaeetvdlvhknladsqahlqgeplnvdslped mrkmrltnapsere...

[[3]]

[1] "mskifvnpsairagladlemaeetvdlinrniednqahlqgepievdnlpedm grlhlddgkspnp...

[[4]]

[1] "mskslihpsdlragladiemadetvdlvyknlsegqahlqgepfdik dlpegvsklqisdnvrsdt...

Only the first part of the alignment stored in *virusaln$seq* is shown here, as it is very long.

VIEWING A LONG MULTIPLE ALIGNMENT

If you want to view a long multiple alignment, it is convenient to view the multiple alignment in blocks. The **R** function "*printMultipleAlignment()*" below will do this for you:

```
> printMultipleAlignment <- function(alignment, chunksize=60)
 {
   # this function requires the Biostrings package
   require("Biostrings")
   # find the number of sequences in the alignment
   numseqs <- alignment$nb
   # find the length of the alignment
   alignmentlen <- nchar(alignment$seq[[1]])
   starts <- seq(1, alignmentlen, by=chunksize)
   n <- length(starts)
   # get the alignment for each of the sequences:
   aln <- vector()
   lettersprinted <- vector()
   for (j in 1:numseqs)
   {
     alignmentj <- alignment$seq[[j]]
     aln[j] <- alignmentj
     lettersprinted[j] <- 0
   }
   # print out the alignment in blocks of 'chunksize' columns:
   for (i in 1:n) { # for each of n chunks
```

```
    for (j in 1:numseqs)
    {
      alnj <- aln[j]
      chunkseqjaln <- substring(alnj, starts[i], starts[i]+chunksize-1)
      chunkseqjaln <- toupper(chunkseqjaln)
      # Find out how many gaps there are in chunkseqjaln:
       gapsj <- countPattern("-",chunkseqjaln) # countPattern() is from
Biostrings package
         # Calculate how many residues of the first sequence we have
printed so far in the alignment:
      lettersprinted[j] <- lettersprinted[j] + chunksize - gapsj
      print(paste(chunkseqjaln,lettersprinted[j]))
    }
    print(paste(' '))
  }
}
```

As its inputs, the function "*printMultipleAlignment()*" takes the input alignment, and the number of columns to print out in each block. For example, to print out the multiple alignment of virus phosphoproteins (which we stored in variable *virusaln*, see above) in blocks of 60 columns, we type:

```
> print Multiple Alignment (virusaln, 60)
    [1] "MSKDLVHPSLIRAGIVELEMAEETTDLINRTIESNQAHL-
QGEPLYVDSLPEDMSRLRIED 60"
  [1] "MSKGLIHPSAIRSGLVDLEMAEETVDLVHKNLADSQAHLQ-
GEPLNVDSLPEDMRKMRLTN 60"
   [1] "MSKIFVNPSAIRAGLADLEMAEETVDLINRNIEDNQAHLQ-
GEPIEVDNLPEDMGRLHLDD 60"
  [1] "MSKSLIHPSDLRAGLADIEMADETVDLVYKNLSEGQAHLQ-
GEPFDIKDLPEGVSKLQISD 60"
  [1] " "
```

[1] "KSRRTKTEEEERDEGSSEEDNYLSEGQDPLIPFQNFLDEI GA-RAVKRLKTGEGFFRVWSA 120"

[1] "APSEREIIEEDEEEYSSEDEYYLSQGQDPMVPFQNFLDEL GT-QIVRRMKSGDGFFKIWSA 120"

[1] "GKSPNPGEMAKVGEGKYREDFQMDEGEDPSLLFQSYLDN VGVQ-IVRQIRSGERFLKIWSQ 120"

[1] "NVRSDTSPNEYSDEDDEEGEDEYEEVYDPVSAFQDFLDET GSYLI-SKLKKGEKIKKTWSE 120"

[1] " "

[1] "LSDDIKGYVSTNIMTSGERDTKSIQIQTEPTASVSSGNESRHD SE-MHDPNDKKDHTPD 179"

[1] "ASEDIKGYVLSTFMKPETQATVSKPTQTDSLSVPRPSQGYTS VPR-DKPSNSESQGGGVK 179"

[1] "TVEEIISYVAVNFP-NPPGKSSEDKSTQTTGRELKKETTPTPS QRESQS-SKARMAAQTAS 179"

[1] "VSRVIYSYVMSNFPPRPPKPTTKDIAVQADLKKPNEIQKISE HKSKS-EPSPREPVVEMHK 180"

[1] " "

[1] "HDVVPDIESSTDKGEIRDIEGEVAHQVAESFSKKYKFPSRSS GIFLWN-FEQLKMNLDDIV 239"

[1] "PKKVQKSEWTRDTDEISDIEGEVAHQVAESFSKKYKFPSRS SGIFL-WNFEQLKMNLDDIV 239"

[1] "GPPALEWSATNEEDDLSVEAEIAHQIAESFSKKYKFPSRSSGI LLYNFE-QLKMNLDDIV 238"

[1] "HATLE-----NPEDDEGALESEIAHQVAESYSKKYKFPSKSSGI FLWNFEQL-KMNLDDIV 235"

[1] " "

[1] "KAAMNVPGVERIAEKGGKLPLRCILGFVALDSSKRFRLLA DNDKVARL-IQEDINSYMARL 299"

[1] "KTSMNVPGVDKIAEKGGKLPLRCILGFVSLDSSKRFRLLAD TDKVARLM-QDDIHNYMTRI 299"

[1] "KEAKNVPGVTRLARDGSKLPLRCVLGWVALANSKKFQLL VESNKLS-KIMQDDLNRYTSC- 297"

```
[1] "QVARGVPGISQIVERGGKLPLRCMLGYVGLETSKRFRSL
VN-QDKLCKLMQEDLNAYSVSS 295"
[1] " "
[1] "EEAE-- 357"
[1] "EEIDHN 359"
[1] "------ 351"
[1] "NN---- 351"
[1] " "
```

DISCARDING VERY POORLY CONSERVED REGIONS FROM AN ALIGNMENT

It is often a good idea to discard very poorly conserved regions from a multiple alignment before building a phylogenetic tree, as the very poorly conserved regions are likely to be regions that are either not homologous between the sequences being considered (and so do not add any phylogenetic signal), or are homologous but are so diverged that they are very difficult to align accurately (and so may add noise to the phylogenetic analysis, and decrease the accuracy of the inferred tree). To discard very poorly conserved regions from a multiple alignment, you can use the following **R** function, "*cleanAlignment()*":

```
> cleanAlignment <- function(alignment, minpcnongap, minpcid)
  {
    # make a copy of the alignment to store the new alignment in:
    newalignment <- alignment
    # find the number of sequences in the alignment
    numseqs <- alignment$nb
    # empty the alignment in "newalignment")
    for (j in 1:numseqs) { newalignment$seq[[j]] <- "" }
    # find the length of the alignment
    alignmentlen <- nchar(alignment$seq[[1]])
    # look at each column of the alignment in turn:
```

```
for (i in 1:alignmentlen)
{
  # see what percent of the letters in this column are non-gaps:
  nongap <- 0
  for (j in 1:numseqs)
  {
    seqj <- alignment$seq[[j]]
    letterij <- substr(seqj,i,i)
    if (letterij != "-") { nongap <- nongap + 1}
  }
  pcnongap <- (nongap*100)/numseqs
  # Only consider this column if at least minpcnongap % of the letters
are not gaps:
  if (pcnongap >= minpcnongap)
  {
  # see what percent of the pairs of letters in this column are identical:
    numpairs <- 0; numid <- 0
    # find the letters in all of the sequences in this column:
    for (j in 1:(numseqs-1))
    {
      seqj <- alignment$seq[[j]]
      letterij <- substr(seqj,i,i)
      for (k in (j+1):numseqs)
      {
        seqk <- alignment$seq[[k]]
        letterkj <- substr(seqk,i,i)
        if (letterij != "-" && letterkj != "-")
        {
          numpairs <- numpairs + 1
          if (letterij == letterkj) { numid <- numid + 1}
```

```
        }
      }
    }
    pcid <- (numid*100)/(numpairs)
    # Only consider this column if at least %minpcid of the pairs of letters are identical:
    if (pcid >= minpcid)
    {
      for (j in 1:numseqs)
      {
        seqj <- alignment$seq[[j]]
        letterij <- substr(seqj,i,i)
        newalignmentj <- newalignment$seq[[j]]
        newalignmentj <- paste(newalignmentj,letterij,sep="")
        newalignment$seq[[j]] <- newalignmentj
      }
    }
  }
}
return(newalignment)
}
```

The function *cleanAlignment()* takes three arguments (inputs): the input alignment; the minimum percent of letters in an alignment column that must be non-gap characters for the column to be kept; and the minimum percent of pairs of letters in an alignment column that must be identical for the column to be kept.

For example, if we have a column with letters "T", "A", "T", "-" (in four sequences), then 75% of the letters are non-gap characters; and the pairs of letters are "T,A", "T,T", and "A,T", and 33% of the pairs of letters are identical.

We can use the function *cleanAlignment()* to discard the very poorly aligned columns from a multiple alignment. For example, if you look at the multiple alignment for the virus phosphoprotein sequences (which we printed out using function *printMultipleAlignment()*, see above), we can see that the last few columns are poorly aligned (contain many gaps and mismatches), and probably add noise to the phylogenetic analysis. Therefore, to filter out the well conserved columns of the alignment, and discard the very poorly conserved columns, we can type:

> cleanedvirusaln <- cleanAlignment(virusaln, 30, 30)

In this case, we required that at least 30% of letters in a column are not gap characters for that column to be kept, and that at least 30% of pairs of letters in an alignment column must be identical for the column to be kept. We can print out the filtered alignment by typing:

> printMultipleAlignment(cleanedvirusaln)

[1] "MSKLVHPSIRAGIVELEMAEETTDLIRTIQAHLQGEPVDLPEDMRLIDR-EEEDEGDPFQF 60"

[1] "MSKLIHPSIRSGLVDLEMAEETVDLVKNLQAHLQGEPVDLPEDMKMLN-SEEEEQGDPFQF 60"

[1] "MSKFVNPSIRAGLADLEMAEETVDLIRNIQAHLQGEPVDLPEDMRLLDS-AERDEGDPFQY 60"

[1] "MSKLIHPSLRAGLADIEMADETVDLVKNLQAHLQGEPIKLPEGVKLIDRE-EEEEVDPFQF 60"

[1] " "

[1] "LDEGVKGEFRWSSIGYVNIMSTSIQTHSDESGEDEEVAHQVAESFSKKY-KFPSRSSGIFL 120"

[1] "LDEGVKGDFKWSSIGYVTFMPTSKQTSDSETDEDEEVAHQVAESFSKKY-FPSRSSGIFL 120"

[1] "LDNGVRGEFKWSVISYVNFPPSDKQTSSSSTDD-EEIAHQIAESFSKKYK-FPSRSSGILL 119"

[1] "LDEGIKGEIKWSSISYVNFPPTDIQAHSS--DDAEEIAHQVAESYSKKYKF-PSKSSGIFL 118"

[1] " "

[1] "WNFEQLKMNLDDIVKANVPGVIAEGGKLPLRCLGVLSKRFRLLADKVRL-IQEDINYEE 180"

[1] "WNFEQLKMNLDDIVKSNVPGVIAEGGKLPLRCLGVLSKRFRLLADKVRLM-QDDIHYEE 180"

[1] "YNFEQLKMNLDDIVKANVPGVLARGSKLPLRCLGVLSKKF QLLVNKLKI-MQDDLNY-- 177"

[1] "WNFEQLKMNLDDIVQAGVPGIIVEGGKLPLRCLGVLSKRF RSLVDKL-KLMQEDLNYNN 178"

[1] " "

The filtered alignment is shorter, but is missing some of the poorly conserved regions of the original alignment. Note that it is not a good idea to filter out too much of your alignment, as if you are left with few columns in your filtered alignment, you will be basing your phylogenetic tree upon a very short alignment (little data), and so the tree may be unreliable. Therefore, you need to achieve a balance between discarding the dodgy (poorly aligned) parts of your alignment, and retaining enough columns of the alignment that you will have enough data to based your tree upon.

CALCULATING GENETIC DISTANCES BETWEEN PROTEIN SEQUENCES

A common first step in performing a phylogenetic analysis is to calculate the pairwise genetic distances between sequences. The genetic distance is an estimate of the divergence between two sequences, and is usually measured in quantity of evolutionary change (an estimate of the number of mutations that have occurred since the two sequences shared a common ancestor).

We can calculate the genetic distances between protein sequences using the "*dist.alignment()*" function in the **SeqinR** package. The *dist. alignment()* function takes a multiple alignment as input. Based on the multiple alignment that you give it, *dist.alignment()* calculates the genetic distance between each pair of proteins in the multiple alignment. For example, to calculate genetic distances between the virus phosphoproteins based on the multiple sequence alignment stored in *virusaln*, we type:

```
> virusdist <- dist.alignment(virusaln)          # Calculate the
genetic distances
> virusdist                                      # Print out the genetic
distance matrix
```

```
         P0C569    O56773    P06747
O56773   0.4142670
P06747   0.4678196 0.4714045
Q5VKP1   0.4828127 0.5067117 0.5034130
```

The genetic distance matrix above shows the genetic distance between each pair of proteins. The sequences are referred to by their UniProt accessions. If you remember from above, P06747 is rabies virus phosphoprotein, P0C569 is Mokola virus phosphoprotein, O56773 is Lagos bat virus phosphoprotein and Q5VKP1 is Western Caucasian bat virus phosphoprotein. Based on the genetic distance matrix above, we can see that the genetic distance between Lagos bat virus phosphoprotein (O56773) and Mokola virus phosphoprotein (P0C569) is smallest (about 0.414). Similarly, the genetic distance between Western Caucasian bat virus phosphoprotein (Q5VKP1) and Lagos bat virus phosphoprotein (O56773) is the biggest (about 0.507). The larger the genetic distance between two sequences, the more amino acid changes or *indels* that have occurred since they shared a common ancestor, and the longer ago their common ancestor probably lived.

CALCULATING GENETIC DISTANCES BETWEEN DNA/MRNA SEQUENCES

Just like for protein sequences, you can calculate genetic distances between DNA (or mRNA) sequences based on an alignment of the sequences. For example, the NCBI accession AF049118 contains mRNA sequence for Mokola virus phosphoprotein, RefSeq AF049114 contains mRNA sequence for Mokola virus phosphoprotein, and AF049119 contains the mRNA sequence for Lagos bat virus phosphoprotein, while AF049115 contains the mRNA sequence for Duvenhage virus phosphoprotein.

To retrieve these sequences from the NCBI database, we can search the ACNUC "genbank" sub-database (since these are nucleotide sequences), by typing:

```
> seqnames <- c("AF049118", "AF049114", "AF049119", "AF049115")
# Make a vector containing the names of the sequences
> seqs <- retrieveseqs(seqnames,"genbank")                # Retrieve the
```

sequences and store them in list variable "seqs"

We can then write out the sequences to a FASTA-format file by typing:

> write.fasta(seqs, seqnames, file="virusmRNA.fasta")

We can then use **CLUSTAL** to create a PHYLIP-format alignment of the sequences, and store it in the alignment file "virusmRNA.phy." This picture shows part of the alignment:

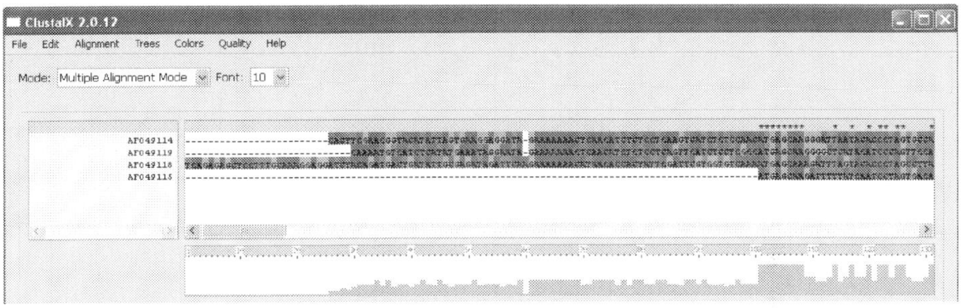

We can then read the alignment into **R**:

> virusmRNAaln <- read.alignment(file = "virusmRNA.phy", format = "phylip")

We saw above that the function *dist.alignment()* can be used to calculate a genetic distance matrix based on a protein sequence alignment.

You can calculate a genetic distance for DNA or mRNA sequences using the dist.dna() function in the **Ape R** package. *dist.dna()*takes a multiple alignment of DNA or mRNA sequences as its input, and calculates the genetic distance between each pair of DNA sequences in the multiple alignment.

The *dist.dna()* function requires the input alignment to be in a special format known as "**DNAbin**" format, so we must use the *as.DNAbin()* function to convert our DNA alignment into this format before using the *dist.dna()* function. For example, to calculate the genetic distance between each pair of mRNA sequences for the virus phosphoproteins, we type:

> virusmRNAalnbin <- as.DNAbin(virusmRNAaln) # Convert the alignment to "DNAbin" format

> virusmRNAdist <- dist.dna(virusmRNAalnbin) # Calculate the genetic distance matrix

> virusmRNAdist # Print out the genetic distance matrix

```
         AF049114 AF049119 AF049118
AF049119 0.3400576
AF049118 0.5235850 0.5637372
AF049115 0.6854129 0.6852311 0.7656023
```

BUILDING AN UNROOTED PHYLOGENETIC TREE FOR PROTEIN SEQUENCES

Once we have a distance matrix that gives the pairwise distances between all our protein sequences, we can build a phylogenetic tree based on that distance matrix. One method for using this is the *neighbour-joining algorithm*. You can build a phylogenetic tree using the neighbour-joining algorithm with the **Ape R** package. First you will need to install the "**ape**" package (see instructions on how to install R packages). It is left to the reader to install and invoke the ape package for executing the code below.

The following **R** function "*unrootedNJtree()*" builds a phylogenetic tree based on an alignment of sequences, using the neighbour-joining algorithm, using functions from the "*ape*" package.

The "*unrootedNJtree()*" function takes an alignment of sequences its input, calculates pairwise distances between the sequences based on the alignment, and then builds a phylogenetic tree based on the pairwise distances. It returns the phylogenetic tree, and also makes a picture of that tree:

```
> unrootedNJtree <- function(alignment,type)
  {
    # this function requires the ape and seqinR packages:
    require("ape")
    require("seqinr")
    # define a function for making a tree:
    makemytree <- function(alignmentmat)
    {
      alignment <- ape::as.alignment(alignmentmat)
```

```
    if   (type == "protein")
    {
      mydist <- dist.alignment(alignment)
    }
    else if (type == "DNA")
    {
      alignmentbin <- as.DNAbin(alignment)
      mydist <- dist.dna(alignmentbin)
    }
    mytree <- nj(mydist)
     mytree <- makeLabel(mytree, space="") # get rid of spaces in tip
names.
    return(mytree)
  }
  # infer a tree
  mymat  <- as.matrix.alignment(alignment)
  mytree <- makemytree(mymat)
  # bootstrap the tree
  myboot <- boot.phylo(mytree, mymat, makemytree)
  # plot the tree:
  plot.phylo(mytree,type="u")  # plot the unrooted phylogenetic tree
  nodelabels(myboot,cex=0.7)   # plot the bootstrap values
   mytree$node.label <- myboot   # make the bootstrap values be the
node labels
  return(mytree)
}
```

To use the function to make a phylogenetic tree, you must first copy and paste the function into **R**. You can then use it to make a tree, for example of the virus phosphoproteins, based on the sequence alignment:

```
> virusalntree <- unrootedNJtree(virusaln,type="protein")
```

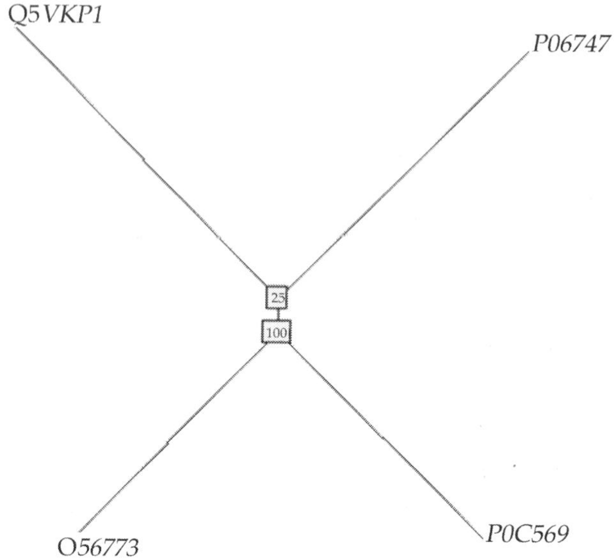

Note that you need to specify that the type of sequences that you are using are protein sequences when you use *unrootedNJtree()*, by setting "*type=protein*".

We can see that Q5VKP1 (Western Caucasian bat virus phosphoprotein) and P06747 (rabies virus phosphoprotein) have been grouped together in the tree, and that O56773 (Lagos bat virus phosphoprotein) and P0C569 (Mokola virus phosphoprotein) are grouped together in the tree.

This is consistent with what we saw above in the genetic distance matrix, which showed that the genetic distance between Lagos bat virus phosphoprotein (O56773) and Mokola virus phosphoprotein (P0C569) is relatively small. The numbers in blue boxes are *bootstrap values* for the nodes in the tree.

A *bootstrap* value for a particular node in the tree gives an idea of the confidence that we have in the *clade* (group) defined by that node in the tree. If a node has a high bootstrap value (near 100%) then we are very confident that the *clade* defined by the node is correct, while if it has a low bootstrap value (near 0%) then we are not so confident. Note that the fact that a bootstrap value for a node is high does not necessarily guarantee that the *clade* defined by the node is correct, but just tells us that it is quite likely that it is correct.

How are these bootstrap values calculated?

The bootstrap values are calculated by making many (for example, 100) random "resamples" of the alignment that the phylogenetic tree was based upon. Each "resample" of the alignment consists of a certain number *x* (e.g., 200) of randomly sampled columns from the alignment. Each "resample" of the alignment (e.g., 200 randomly sampled columns) forms a sort of fake alignment of its own, and a phylogenetic tree can be based upon the "resample." We can make 100 random resamples of the alignment, and build 100 phylogenetic trees based on the 100 resamples. These 100 trees are known as the "bootstrap trees." For each *clade* (grouping) that we see in our original phylogenetic tree, we can count in how many of the 100 bootstrap trees it appears. This is known as the "bootstrap value" for the *clade* in our original phylogenetic tree.

For example, if we calculate 100 random resamples of the virus phosphoprotein alignment, and build 100 phylogenetic trees based on these resamples, we can calculate the bootstrap values for each *clade* in the virus phosphoprotein phylogenetic tree.

In this case, the bootstrap value for the node defining the *clade* containing Q5VKP1 (Western Caucasian bat virus phosphoprotein) and P06747 (rabies virus phosphoprotein) is 25%, while the bootstrap value for node defining the clade containg of Lagos bat virus phosphoprotein (O56773) and Mokola virus phosphoprotein (P0C569) is 100%. The bootstrap values for each of these *clades* is the percent of 100 bootstrap trees that the *clade* appears in.

Therefore, we are very confident that Lagos bat virus and Mokola virus phosphoproteins should be grouped together in the tree. However, we are not so confident that the Western Caucasian bat virus and rabies virus phosphoproteins should be grouped together.

The lengths of the branches in the plot of the tree are proportional to the amount of evolutionary change (estimated number of mutations) along the branches. In this case, the branches leading to Lagos bat virus phosphoprotein (O56773) and Mokola virus phosphoprotein (P0C569) from the node representing their common ancestor are slightly shorter than the branches leading to the Western Caucasian bat virus (Q5VKP1) and rabies virus (P06747) phosphoproteins from the node representing their common ancestor.

Let us see what this means. This suggests that there might have been more mutations in the Western Caucasian bat virus (Q5VKP1) and rabies virus (P06747) phosphoproteins since they shared a common ancestor, than in the Lagos bat virus phosphoprotein (O56773) and Mokola virus phosphoprotein (P0C569) since they shared a common ancestor.

The tree above of the virus phosphoproteins is an *unrooted* phylogenetic tree as it does not contain an *outgroup* sequence, that is a sequence of a protein that is known to be more distantly related to the other proteins in the tree than they are to each other. As a result, we cannot tell which direction evolutionary time ran in along the internal branches of the tree. For example, we cannot tell whether the node representing the common ancestor of (O56773, P0C569) was an ancestor of the node representing the common ancestor of (Q5VKP1, P06747), or the other way around.

In order to build a *rooted* phylogenetic tree, we need to have an outgroup sequence in our tree. In the case of the virus phosphoproteins, this is unfortunately not possible, as there is not any protein known that is more distantly related to the four proteins already in our tree than they are to each other. However, in many other cases, an outgroup - a sequence known to be more distantly related to the other sequences in the tree than they are to each other - is known. Hence, it is possible to build a *rooted* phylogenetic tree.

We discussed above that it is a good idea to investigate whether discarding the poorly conserved regions of a multiple alignment has an effect on the phylogenetic analysis. In this case, we made a filtered copy of the multiple alignment and stored it in the variable *cleanedvirusaln* (see above). We can make a phylogenetic tree based on this filtered alignment, and see if it agrees with the phylogenetic tree based on the original alignment:

```
> cleanedvirusalntree <- unrootedNJtree(cleanedvirusaln,type=”prote
in”)
```

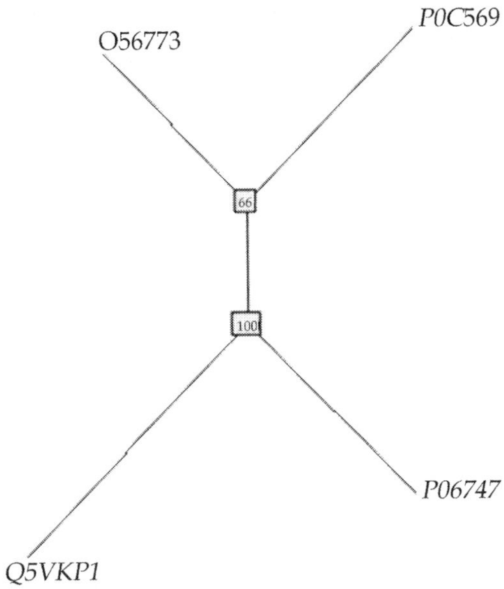

Here O56773 and P0C569 are grouped together, and Q5VKP1 and P06747 are grouped together, as in the phylogenetic tree based on the raw (unfiltered) multiple alignment (see above). Thus, filtering the multiple alignment does not have an effect on the tree.

If we had found a difference in the trees made using the unfiltered and filtered multiple alignments, we would have to examine the multiple alignments closely, to see if the unfiltered multiple alignment contains a lot of very poorly aligned regions that might be adding noise to the phylogenetic analysis (if this is true, the tree based on the filtered alignment is likely to be more reliable).

BUILDING A ROOTED PHYLOGENETIC TREE FOR PROTEIN SEQUENCES

In order to convert the *unrooted* tree into a *rooted* tree, we need to add an outgroup sequence. Normally, the outgroup sequence is a sequence that we know from some prior knowledge to be more distantly related to the other sequences under study than they are to each other. For example, the protein Fox-1 is involved in determining the sex (gender) of an embryo in the nematode worm *Caenorhabditis elegans* (UniProt accession

Q10572). Related proteins are found in other nematodes, including *Caenorhabditis remanei* (UniProt E3M2K8), *Caenorhabditis briggsae* (A8WS01), *Loa loa* (E1FUV2), and *Brugia malayi* (UniProt A8NSK3).

Note that *Caenorhabditis elegans* is a model organism commonly studied in molecular biology. The nematodes *Loa loa*, and *Brugia malayi* are parasitic nematodes that cause filariasis, which is classified by the WHO as a neglected tropical disease.

The UniProt database contains a distantly related sequence from the fruitfly *Drosophila melanogaster* (UniProt accession Q9VT99). If we were to build a phylogenetic tree of the nematode worm Fox-1 homologues, the distantly related sequence from fruitfly would probably be a good choice of outgroup, since the protein is from a different animal group (insects) than the nematode worms. Thus, it is likely that the fruitfly protein is more distantly related to all the nematode proteins than they are to each other. To retrieve the sequences from UniProt we can use the "retrieveseqs()" function (see above):

```
> seqnames <- c(“Q10572”,”E3M2K8”,”Q8WS01”,”E1FUV2”,”A8NSK3”,”Q9VT99”)
```

```
> seqs <- retrieveseqs(seqnames,”swissprot”)
```

We can then write out the sequences to a FASTA file:

```
> write.fasta(seqs, seqnames, file=”fox1.fasta”)
```

We can then use CLUSTAL to create a PHYLIP-format alignment of the sequences, and store it in the alignment file "fox1.phy." This picture shows part of the alignment (the alignment is quite long, so not all of it is shown):

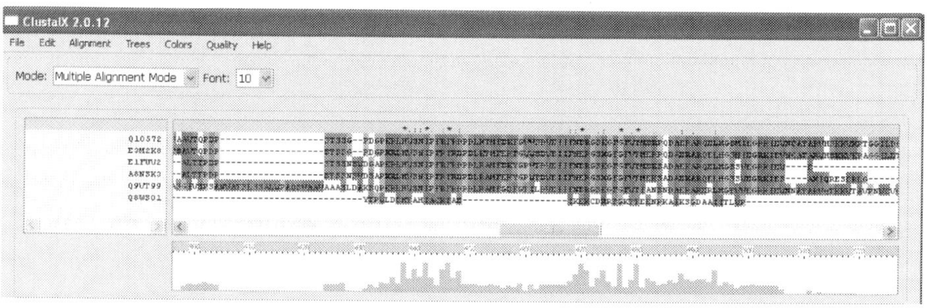

We can then read the alignment into **R**:

```
> fox1aln <- read.alignment(file = “fox1.phy”, format = “phylip”)
```

The next step is to build a phylogenetic tree of the proteins, which again we can do using the neighbour-joining algorithm. This time we have an outgroup in our set of sequences, so we can build a rooted tree. The function "*rootedNJtree()*" can be used to build a rooted tree. It returns the phylogenetic tree, and also makes a picture of the tree:

```
> rootedNJtree <- function(alignment, theoutgroup, type)
 {
   # load the ape and seqinR packages:
   require("ape")
   require("seqinr")
   # define a function for making a tree:
   makemytree <- function(alignmentmat, outgroup=`theoutgroup`)
   {
     alignment <- ape::as.alignment(alignmentmat)
     if    (type == "protein")
     {
       mydist <- dist.alignment(alignment)
     }
     else if (type == "DNA")
     {
       alignmentbin <- as.DNAbin(alignment)
       mydist <- dist.dna(alignmentbin)
     }
     mytree <- nj(mydist)
      mytree <- makeLabel(mytree, space="") # get rid of spaces in tip
names.
     myrootedtree <- root(mytree, outgroup, r=TRUE)
     return(myrootedtree)
   }
   # infer a tree
   mymat  <- as.matrix.alignment(alignment)
```

```
    myrootedtree <- makemytree(mymat, outgroup=theoutgroup)
    # bootstrap the tree
    myboot <- boot.phylo(myrootedtree, mymat, makemytree)
    # plot the tree:
    plot.phylo(myrootedtree, type="p")  # plot the rooted phylogenetic tree
    nodelabels(myboot,cex=0.7)         # plot the bootstrap values
    mytree$node.label <- myboot   # make the bootstrap values be the node labels
    return(mytree)
  }
```

The function takes the alignment and the name of the outgroup as its inputs. For example, to use it to make a phylogenetic tree of the *C. elegans* Fox-1 protein and its homologues, using the fruitfly protein (UniProt Q9VT99) as the outgroup, we type:

```
> fox1alntree <- rootedNJtree(fox1aln, "Q9VT99",type="protein")
```

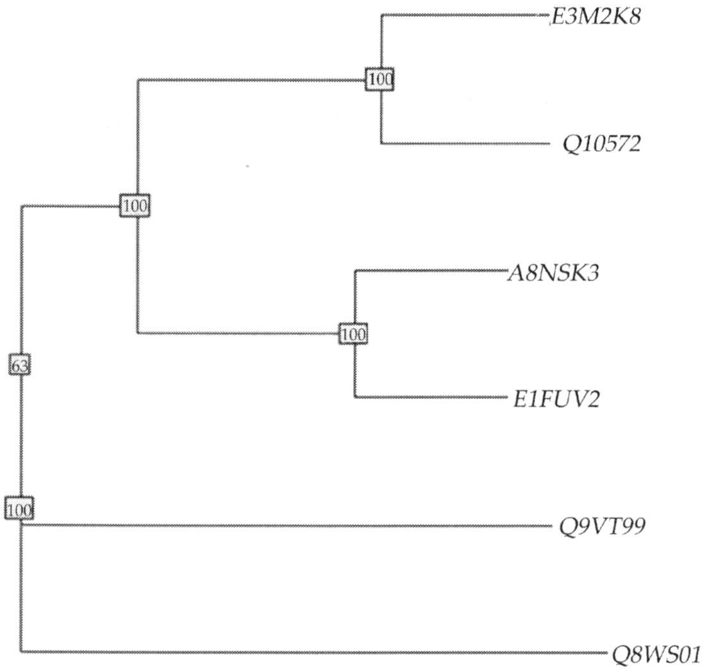

Here we can see that E3M2K8 (*C. remanei* Fox-1 homologue) and Q10572 (*C. elegans* Fox-1) have been grouped together with bootstrap 100%, and A8NSK3 (*Brugia malayi* Fox-1 homologue) and E1FUV2 (*Loa loa* Fox-1 homologue) have been grouped together with bootstrap 100%. These four proteins have also been grouped together in a larger *clade* with bootstrap 100%. Compared to these four proteins, the Q8WS01 (*C. briggsae* Fox-1 homologue) and Q9VT99 (fruitfly outgroup) seem to be relatively distantly related. As this is a *rooted* tree, we know the direction that evolutionary time ran. Say we call the ancestor of the four sequences (E3M2K8, Q10572, A8NSK3, E1FUV2) *ancestor1*, the ancestor of the two sequences (E3M2K8, Q10572) *ancestor2*, and the ancestor of the two sequences (A8NSK3, E1FUV2) *ancestor3*. Because it is a *rooted* tree, we know that time ran from left to right along the branches of the tree, so that *ancestor1* was the ancestor of *ancestor2*, and *ancestor1* was also the ancestor of *ancestor3*. In other words, *ancestor1* lived before *ancestor2* or *ancestor3*; *ancestor2* and *ancestor3* were descendants of *ancestor1*. Another way of saying this is that E3M2K8 and Q10572 shared a common ancestor with each other more recently than they did with A8NSK3 and E1FUV2.

The lengths of branches in this tree are proportional to the amount of evolutionary change (estimated number of mutations) that occurred along the branches. The branches leading back from E3M2K8 and Q10572 to their last common ancestor are slightly longer than the branches leading back from A8NSK3 and E1FUV2 to their last common ancestor. This indicates that there has been more evolutionary change in E3M2K8 (*C. remanei* Fox-1 homologue) and Q10572 (*C. elegans* Fox-1) proteins since they diverged, than there has been in A8NSK3 (*Brugia malayi* Fox-1 homologue) and E1FUV2 (*Loa loa* Fox-1 homologue) since they diverged.

BUILDING A PHYLOGENETIC TREE FOR DNA OR MRNA SEQUENCES

In the example above, a phylogenetic tree was built for protein sequences. The genomes of distantly related organisms such as vertebrates will have accumulated many mutations since they diverged. Sometimes, so many mutations have occurred since the organisms diverged that their DNA sequences are hard to align correctly and it is also hard to accurately estimate evolutionary distances from alignments of those DNA sequences.

In contrast, as many mutations at the DNA level are synonymous at the protein level, protein sequences diverge at a slower rate than DNA sequences. This is why for reasonably distantly related organisms such as vertebrates, *it is usually preferable to use protein sequences for phylogenetic analyses*. If you are studying closely related organisms such as primates, few mutations will have occurred since they diverged. As a result, if you use protein sequences for a phylogenetic analysis, there may be too few amino acid substitutions to provide enough 'signal' to use for the phylogenetic analysis. Therefore, it is often preferable to use DNA sequences for a phylogenetic analysis of closely related organisms such as primates.

We can use the functions *unrootedNJtree()* and *rootedNJtree()* described above to build *unrooted* or *rooted* neighbour-joining phylogenetic trees based on an alignment of DNA or mRNA sequences. In this case, we need to use "*type=DNA*" as an argument in these functions, to tell them that we are making a tree of nucleotide sequences, not protein sequences.

For example, to build an *unrooted* phylogenetic tree based on the alignment of the virus phosphoprotein mRNA sequences, we type in **R**:

```
> virusmRNAaln <- read.alignment(file = "virusmRNA.phy", format = "phylip")
> virusmRNAalntree <- unrootedNJtree(virusmRNAaln, type="DNA")
```

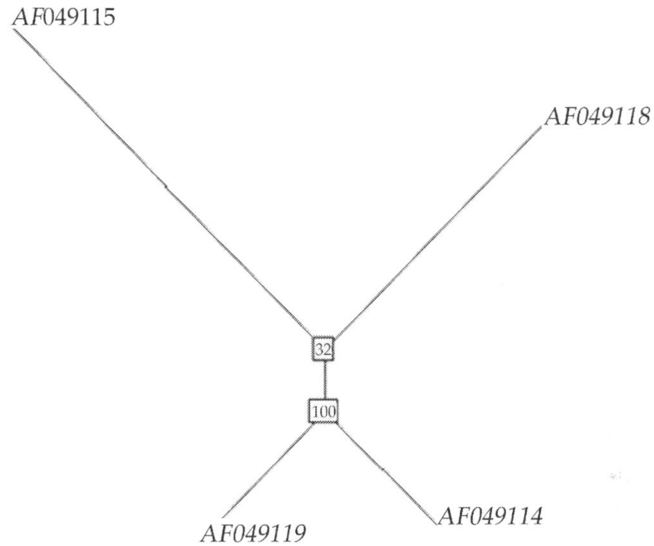

SAVING A PHYLOGENETIC TREE AS A NEWICK-FORMAT TREE FILE

A commonly used format for representing phylogenetic trees is the **Newick** format. Once you have built a phylogenetic tree using **R**, it is convenient to store it as a Newick-format tree file. This can be done using the "*write.tree()*" function in the **Ape R** package.

For example, to save the *unrooted* phylogenetic tree of virus phosphoprotein mRNA sequences as a Newick-format tree file called "*virusmRNA.tre*", we type:

> write.tree(virusmRNAalntree, "virusmRNA.tre")

The Newick-format file "*virusmRNA.tre*" should now appear in your "My Documents" folder.

SUMMARY

In this chapter, you have learnt the following R functions that belong to the bioinformatics packages:

- *read.alignment()* from the **SeqinR** package for reading in a multiple alignment
- *dist.alignment()* from the **SeqinR** package for calculating genetic distances between protein sequences
- *dist.dna()* from the **Ape** package for calculating genetic distances between DNA or mRNA sequences

REFERENCES

1. Little Book of R for Bioinformatics by Avril Coghlan, Wellcome Trust Sanger Institute, Cambridge, U.K (**2016**).
2. Introduction to Computational Genomics: A Case Studies Approach by Cristianini N & Hahn MW (Cambridge University Press, **2006**)
3. Molecular Evolution: A Phylogenetic Approach by Page RDM & Holmes EC (Wiley-Blackwell, **2009**).

7

GENES AND GENOMES

Biologists are faced with a large ocean of sequences in the form of genes. But, do any of these have any functional significane? This involves access to an encyclopedia of vast information in the domain of genes and genomes. This has been made possible by accessing KEGG database.

KEGG

The genome sequencing projects aim at determining gene catalogs for a large number of organisms, functional annotation of individual genes is still largely incomplete. This propmpted the scientific community to direct their efforts towards creating a repository for functional annotation of genes and genomes. The KEGG project was initiated in May 1995 under the Human Genome Program of the Ministry of Education, Science, Sports and Culture in Japan. KEGG (Kyoto Encyclopedia of Genes and Genomes) is an effort to link genomic information with higher order functional information by computerizing current knowledge on cellular processes and by standardizing gene annotations.

KEGG (**K**yoto **E**ncyclopedia of **G**enes and **G**enomes) is a knowledge base for systematic analysis of gene functions, linking genomic information with higher order functional information. The genomic information is stored in the GENES database, which is a collection of gene catalogs for all the completely sequenced genomes and some partial genomes with up-to-date annotation of gene functions. The higher order functional information is stored in the PATHWAY database, which contains graphical representations of cellular processes, such as metabolism, membrane transport, signal transduction and cell cycle. The PATHWAY database is supplemented by a set of ortholog group tables for the information about conserved subpathways (pathway motifs), which are often encoded by positionally coupled genes on the chromosome and which are especially useful in predicting gene functions. A third database in KEGG is LIGAND for the information about chemical compounds, enzyme molecules and enzymatic reactions. KEGG provides Java graphics tools for browsing genome maps, comparing two genome maps and manipulating expression maps, as well as computational tools for sequence comparison, graph comparison and path computation. The KEGG databases are daily updated and made freely available at:

http://www.genome.ad.jp/kegg/

The biological function of the living cell is a result of many interacting molecules at varying timescales; and it cannot be attributed to just a single gene or a single molecule. The functional assignment in KEGG is a process of linking a set of genes in the genome with a network of interacting molecules in the cell, such as a *pathway* or a *complex*, representing a higher order biological function.

KEGG consists of three databases:

1. PATHWAY for representation of higher order functions in terms of the network of interacting molecules,
2. GENES for the collection of gene catalogs for all the completely sequenced genomes and some partial genomes, and
3. LIGAND for the collection of chemical compounds in the cell, enzyme molecules and enzymatic reactions.

This was the architecture of KEGG when Prof. Minoru Kanehisa launched the service. Over years it has grown substantially and presently has the following broad sections:

- Systems information
    - PATHWAY — pathway maps for cellular and organismal functions
    - MODULE — modules or functional units of genes
    - BRITE — hierarchical classifications of biological entities
- Genomic information
    - GENOME — complete genomes
    - GENES — genes and proteins in the complete genomes
    - ORTHOLOGY — ortholog groups of genes in the complete genomes
- Chemical information
    - COMPOUND, GLYCAN — chemical compounds and glycans
    - REACTION, RPAIR, RCLASS — chemical reactions
    - ENZYME — enzyme nomenclature
- Health information
    - DISEASE — human diseases
    - DRUG — approved drugs
    - ENVIRON — crude drugs and health-related substances

We explain these briefly in the following section. A word on KEGG Identifiers. Each database entry in KEGG is identified by the form of 'db:entry' where 'db' is the database name and 'entry' is the entry identifier.

SYSTEMS INFORMATION

The pathway map in KEGG PATHWAY represents our knowledge about various types of molecular networks: reaction/interaction networks for metabolism, genetic information processing, environmental information processing and other cellular processes; perturbed reaction/interaction networks for human diseases; and relation networks (chemical structure transformation networks) for drug development. The molecular network shown in the pathway map is a graph consisting of nodes (orthologs, genes, proteins, small molecules, etc.) and edges (reactions, interactions and relations).

The KEGG reference pathway maps and BRITE reference hierarchies are created in a general way to be applicable to all organisms; namely, in terms of the orthologs defined by *K* numbers. **KEGG BRITE** is a collection of manually created hierarchical text (htext) files capturing functional hierarchies of various biological objects, especially those represented as KEGG objects.

In contrast to KEGG PATHWAY, which is limited to molecular interactions and reactions, KEGG BRITE incorporates many different types of relationships. The hierarchical list in KEGG BRITE represents hierarchically organized membership information, which may be called ontology. It is used to represent our knowledge on functional hierarchies inherent in various types of biological systems; not only molecular systems, but also cellular and organismal systems. The content of KEGG BRITE is currently classified into: genes and proteins; compounds and reactions; drugs and diseases; and cells and organisms.

The organism-specific pathways and hierarchies can then be generated by converting *K* numbers to gene identifiers in a given organism. Since kast decade, the KEGG PATHWAY database has been completely renovated. All the pathway maps have been redrawn using a newly developed tool called KegSketch, which generates KGML+ (meaning KGML + SVG) files. Internally the database update procedure is now based on the text manipulation of these files rather than the color manipulation of image files. For outside services, the coloring procedure continues to be done on image files, but the image file format has been changed from GIF to PNG to accommodate more colors. As a result, there is now no distinction between the global map and the regular pathway maps; they can be manipulated in the same way both in the new KEGG Atlas tool and the traditional image map viewer. Another new feature in the outside service is the XML version of KEGG pathway maps, which is made available in both the original KGML format and the converted BioPAX level 2 format. We will take a hands-on approach on this aspect next in the chapter.

GENOMIC INFORMATION

One of the main objectives of the KEGG project has been to uncover higher level systemic functions of the cell and the organism from genomic and molecular-level information. The basis for genome annotation in KEGG, which is continuously performed for all sequenced genomes, is the KO system consisting of manually defined ortholog groups that correspond to individual nodes in the KEGG pathway maps and the BRITE functional hierarchies. Once genes are assigned KO identifiers or *K* numbers by the ortholog annotation procedure described below, the collective body of *K* numbers can be mapped to KEGG pathway maps and BRITE functional hierarchies, highlighting any subsystems present and enabling higher level functional interpretation of the genome.

The current KEGG annotation procedure is as follows.

- Gene information for completely sequenced genomes is computationally generated from RefSeq and other public resources, and stored in the KEGG GENES database.
- Sequence similarity scores and best-hit relations are computationally generated from KEGG GENES by pair-wise genome comparisons using SSEARCH, and stored in the KEGG SSDB database.
- Automatic genome-based annotation is performed for a limited set (currently, about one-third) of *K* numbers, which are considered safe for such purpose based on the result of SSDB computation and the criteria of the KOALA tool.
- Manual annotation is performed across species for other *K* numbers using the KOALA and GFIT tools. This step may

involve addition/revision of ortholog groups, which is essential to increase the number of safe *K* numbers.

As an end user, you can be sure that you would have correct result for your queries in the database. Still, caution and a keen eye is advised as the database is under contnuous development and the authors would be open to suggestions and improvements.

CHEMICAL INFORMATION

The KEGG metabolic pathway maps are drawn to represent the dual aspects of the metabolic network: the genomic network of how genome-encoded enzymes are connected to catalyze consecutive reactions and the chemical network of how chemical structures of substrates and products are transformed by these reactions. A set of enzyme genes in the genome will identify enzyme relation networks when superimposed on the KEGG pathway maps, which in turn characterize chemical structure transformation networks allowing interpretation of biosynthetic and biodegradation potentials of the organism. Alternatively, a set of metabolites identified in the metabolome will lead to the understanding of enzymatic pathways and enzyme genes involved.

The databases in the chemical information category, which are collectively called KEGG LIGAND, are organized by capturing knowledge of the chemical network. In the beginning of the KEGG project, KEGG LIGAND consisted of three databases: KEGG COMPOUND for chemical compounds, KEGG REACTION for chemical reactions, and KEGG ENZYME for reactions in the enzyme nomenclature. Currently, there are additional databases: KEGG GLYCAN for glycans[8] and two auxiliary reaction databases called RPAIR (reactant pair alignments) and RCLASS (reaction class). KEGG COMPOUND has also been expanded to contain various compounds such as xenobiotics, in addition to metabolites.

HEALTH INFORMATION

In KEGG, diseases are viewed as perturbed states of the biological system caused by perturbants of genetic factors and environmental factors, and drugs are viewed as different types of perturbants. The KEGG PATHWAY database includes not only the normal states but also the perturbed states of the biological systems. However, disease pathway maps cannot be drawn for most diseases because molecular mechanisms are not well understood. An alternative approach is taken in the KEGG DISEASE database, which simply catalogs known genetic factors and environmental factors of diseases. These catalogs may eventually lead to more complete wiring diagrams of diseases.

The KEGG DRUG database contains active ingredients of approved drugs in Japan, the USA, and Europe. They are distinguished by chemical structures and/or chemical components and associated with target molecules, metabolizing enzymes, and other molecular interaction network information in the KEGG pathway maps and the BRITE hierarchies. This enables an integrated analysis of drug interactions with genomic information. Crude drugs and other health-related substances, which are outside of the category of approved drugs, are stored in the KEGG ENVIRON database. The databases in the health information category are collectively called KEGG MEDICUS, which also includes package inserts of all marketed drugs in Japan.

With this introduction, we now step into computationally accessing and manipulating information from KEGG database. The readers are suggested to carry this out in IPython notebook.

ACCESSING KEGG

This is an introduction to the Bio.KEGG and Bio.Graphics.KGML modules, providing examples of usage, including data retrieval, manipulation and output rendering.

We need to import Biopython and parts of the KEGG and KGML modules, but first we'll check that a suitably recent version of Biopython is installed (1.65 or higher for Python 2; 1.67 or higher for Python 3):

import Bio Bio.__version__

'1.68'

We're going to be using some other Biopython modules, and working with images in the notebook, so we also prepare for this, with the following imports.

```
from Bio import SeqIO
from Bio.KEGG.REST import *
from Bio.KEGG.KGML import KGML_parser
from Bio.Graphics.KGML_vis import KGMLCanvas
from Bio.Graphics.ColorSpiral import ColorSpiral
from IPython.display import Image, HTML
import os
 import random
```

To visualise the PDF output generated by this library inline, and to provide a head-like peek at text output, the functions below are defined:

```
# A bit of code that will help us display the PDF output
def PDF(filename):
        return HTML('<iframe src=%s width=700 height=350></
iframe>' % filename)
# A bit of helper code to shorten long text
def head(text, lines=10):
 """Print the first lines lines of the passed text."""
        print('\n'.join(text.split('\n')[:lines] + ['[...]']))
```

RETRIEVING DATA FROM KEGG WITH BIO.KEGG.REST

The KEGG database provides a RESTful interface to its data. For more information check the following weblink:

http://www.kegg.jp/kegg/rest/

which we can use via Biopython's Bio.KEGG.REST module.

The KEGG interface is not as well documented as some other resources (such as NCBI or Ensembl), and KEGG does not provide any usage guidelines. To avoid risking overloading the service, Biopython restricts us to three calls per second. Be warned also that the conditions of service include:

"This service should not be used for bulk data downloads".

Biopython's interface converts text data to String handles, but leaves image data as the Byte type.

Obtaining information about a database with kegg_info()

We can obtain basic information about the KEGG databases (e.g. "pathway", "kegg", "hsa", etc.) using the *kegg_info()* function. This returns a handle which can be read as plaintext, with some useful information about the resource.

```
# Information about the Kyoto Encyclopedia of Genes and Genomes
print(kegg_info("kegg").read())
kegg     Kyoto Encyclopedia of Genes and Genomes kegg     Release
82.0+/05-16, May 17          Kanehisa Laboratories        pathway
506,543 entries          brite     183,525 entries        module
407,030 entries          disease      1,829 entries         drug
10,333 entries           environ      850 entries      orthology
20,957 entries           genome       5,120 entries        genes
22,063,068 entries       compound    17,989 entries       glycan
11,015 entries           reaction    10,543 entries       rclass
3,114 entries            enzyme       6,963 entries

# Information about the KEGG Pathway Database
print(kegg_info("pathway").read())
pathway     KEGG Pathway Database
path   Release 82.0+/05-16, May 17 Kanehisa Laboratories 506,543
entries

# Information about the Escherichia coli K-12 MG1655 KEGG Genes
Database
print(kegg_info("eco").read())
T00007      Escherichia coli K-12 MG1655 KEGG Genes Database
```

eco Release 82.0+/05-14, May 17 Kanehisa Laboratories 4,498 entries

Obtaining a list of entries in a database with kegg_list()

The kegg_list() function retrieves a handle, which provides a plaintext tab-separated key:value list of cntry identifiers and definitions in the given database or set of database entries.

The kegg_list() function allows the organism to be specified for a database, with an optional second argument.

```
# List all pathways in the pathway database
head(kegg_list('pathway').read())
path:map00010	Glycolysis / Gluconeogenesis
path:map00020	Citrate cycle (TCA cycle)
path:map00030	Pentose phosphate pathway
path:map00040	Pentose and glucuronate interconversions
path:map00051	Fructose and mannose metabolism
path:map00052	Galactose metabolism
path:map00053	Ascorbate and aldarate metabolism
path:map00061	Fatty acid biosynthesis
path:map00062	Fatty acid elongation
path:map00071	Fatty acid degradation
[...]
```

Now list pathways for a specific organism

```
# Only list pathways present in E. coli K-12 MG1655
head(kegg_list('pathway', 'eco').read())
path:eco00010	Glycolysis / Gluconeogenesis - Escherichia coli K-12 MG1655
path:eco00020	Citrate cycle (TCA cycle) - Escherichia coli K-12 MG1655
path:eco00030	Pentose phosphate pathway - Escherichia coli K-12 MG1655
```

path:eco00040 Pentose and glucuronate interconversions - Escherichia coli K-12 MG1655

path:eco00051 Fructose and mannose metabolism - Escherichia coli K-12 MG1655 path:eco00052 Galactose metabolism - Escherichia coli K-12 MG1655

path:eco00053 Ascorbate and aldarate metabolism - Escherichia coli K-12 MG1655 path:eco00061 Fatty acid biosynthesis - Escherichia coli K-12 MG1655

path:eco00071 Fatty acid degradation - Escherichia coli K-12 MG1655

path:eco00121 Secondary bile acid biosynthesis - Escherichia coli K-12 MG1655

[…]

If only the organism is specified, then a list of gene entries is returned.

E. coli K-12 MG1655 genes

head(kegg_list('eco').read())

eco:b0001 K08278 thr operon leader peptide | (RefSeq) thrL; thr operon leader peptide

eco:b0002 K12524 bifunctional aspartokinase / homoserine dehydrogenase 1 [EC:2.7.2.4 1.1.1.3] | (RefSeq) thrA; Bifunctional aspartokinase/homoserine dehydrogenase 1

eco:b0003 K00872 homoserine kinase [EC:2.7.1.39] | (RefSeq) thrB; homoserine kinase eco:b0004 K01733 threonine synthase [EC:4.2.3.1] | (RefSeq) thrC; L-threonine synthase

eco:b0005 no KO assigned | (RefSeq) yaaX; DUF2502 family putative periplasmic protein

eco:b0006 K09861 uncharacterized protein | (RefSeq) yaaA; peroxide resistance protein, lowers intracellular iron

eco:b0007 K03310 alanine or glycine:cation symporter, AGCS family | (RefSeq) yaaJ; putative transporter

eco:b0008 K00616 transaldolase [EC:2.2.1.2] | (RefSeq) talB; transaldolase B

eco:b0009 K03831 molybdopterin adenylyltransferase [EC:2.7.7.75] | (RefSeq) mog; molybdochelatase incorporating molybdenum into molybdopterin eco:b0010 K07034 uncharacterized protein | (RefSeq) satP; succinate-acetate transporter

[...]

The KEGG API also allows the + operator to be used to combine list identifiers.

Compound entry C01290 and glycan entry G00092

print(kegg_list('C01290+G00092').read())

cpd:C01290 beta-D-Galactosyl-(1->4)-beta-D-glucosyl-(1<->1)-ceramide; beta-D-Galactosyl-1,4-beta-D-glucosylceramide; Lactosylceramide; Gal-beta1->4Glc-beta1->1'Cer; LacCer; Lactosyl-N-acylsphingosine; D-Galactosyl-1,4-beta-D-glucosylceramide

gl:G00092 Lactosylceramide; CDw17; LacCer; (Gal)1 (Glc)1 (Cer)1

But kegg_list() and other functions will take a Python list of identifiers, and combine them automatically:

Compound entry C01290 and glycan entry G00092

print(kegg_list(['C01290', 'G00092']).read())

cpd:C01290 beta-D-Galactosyl-(1->4)-beta-D-glucosyl-(1<->1)-ceramide; beta-D-Galactosyl-1,4-beta-D-glucosylceramide; Lactosylceramide; Gal-beta1->4Glc-beta1->1'Cer; LacCer; Lactosyl-N-acylsphingosine; D-Galactosyl-1,4-beta-D-glucosylceramide

gl:G00092 Lactosylceramide; CDw17; LacCer; (Gal)1 (Glc)1 (Cer)1

Finding a specific entry with kegg_find()

The kegg_find(database, query) function finds entries with matching query keywords or other query data in a given database. Data is returned as tab-separated key:value plain text.

Find shiga toxin genes

head(kegg_find('genes', 'shiga+toxin').read())

ece:Z1464 K11006 shiga toxin subunit A | (GenBank) stx2A; shiga-like toxin II A subunit encoded by bacteriophage BP-933W ece:Z1465 K11007 shiga toxin subunit B | (GenBank) stx2B; shiga-like toxin

II B subunit encoded by bacteriophage BP-933W

ece:Z3343 K11007 shiga toxin subunit B | (GenBank) stx1B; shiga-like toxin 1 subunit B encoded within prophage CP-933V ece:Z3344 K11006 shiga toxin subunit A | (GenBank) stx1A; shiga-like toxin 1 subunit A encoded within prophage CP-933V

ecs:ECs1205 K11006 shiga toxin subunit A | (RefSeq) Shiga toxin 2 subunit A

ecs:ECs1206 K11007 shiga toxin subunit B | (RefSeq) Shiga toxin 2 subunit B

ecs:ECs2973 K11007 shiga toxin subunit B | (RefSeq) Shiga toxin I subunit B

ecs:ECs2974 K11006 shiga toxin subunit A | (RefSeq) Shiga toxin I subunit A

ecf:ECH74115_2905 K11007 shiga toxin subunit B | (GenBank) shigatoxin 2, subunit B

ecf:ECH74115_2906 K11006 shiga toxin subunit A | (GenBank) shiga toxin subunit A

[…]

Find shiga toxin genes only in Escherichia coli O111 H-11128 (EHEC)

print(kegg_find('eoi', 'shiga+toxin').read())

eoi:ECO111_2429 K11007 shiga toxin subunit B | (GenBank) Shiga toxin 2 subunit B

eoi:ECO111_2430 K11006 shiga toxin subunit A | (GenBank) Shiga toxin 2 subunit A

eoi:ECO111_3361 K11006 shiga toxin subunit A | (GenBank) Shiga toxin 1 subunit A

eoi:ECO111_3362 K11007 shiga toxin subunit B | (GenBank) Shiga toxin 1 subunit B

The API lets us query on useful properties, such as compound molecular weight:

Compounds with molecular weight between 300 and 310g/mol

head(kegg_find('compound', '300-310/mol_weight').read())

cpd:C00051 307.32348

cpd:C00200 306.33696

cpd:C00219 304.46688

```
cpd:C00239 307.197122
cpd:C00270 309.26986
cpd:C00357 301.187702
cpd:C00365 308.181882
cpd:C00389 302.2357
cpd:C00732 308.37276
cpd:C00777 300.43512
[…]
```

Retrieving specific entries with kegg_get()

The kegg_get() function lets us retrieve data from KEGG in a range of different formats.

Database entries can be retrieved directly as plain text; sequences may be returned as database entries, amino acid or nucleotide sequences in FASTA format; compounds as database entries, or images (.gif); pathways as database entries, KGML, or images (.png).

Compounds can be retrieved as their database entry in plain text, or as a .gif image.

 # Compound as database entry

```
head(kegg_get("cpd:C01290").read())
ENTRY       C01290                 Compound
NAME    beta-D-Galactosyl-(1->4)-beta-D-glucosyl-(1<->1)-ceramide; beta-D-Galactosyl-1,4-beta-D-glucosylceramide; Lactosylceramide; Gal-beta1->4Glc-beta1->1'Cer; LacCer; Lactosyl-N-acylsphingosine; D-Galactosyl-1,4-beta-D-glucosylceramide
FORMULA    C31H56NO13R
REMARK     Same as: G00092
[…]
```

 # Compound as image

```
Image(kegg_get("cpd:C01290", "image").read())
```

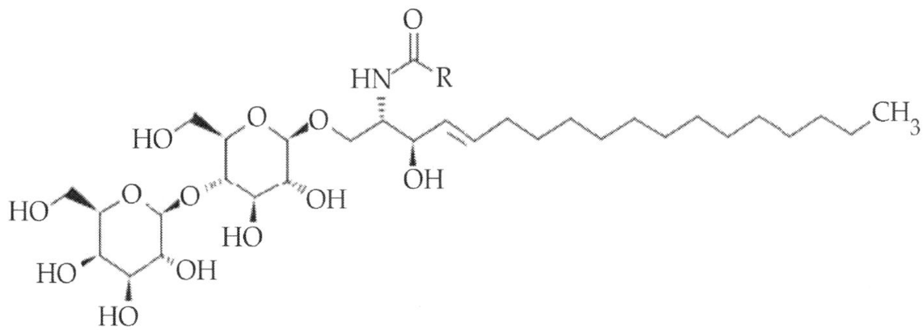

Gene data can be retrieved as the plain text database entry, or a FASTA nucleotide or protein sequence.

 # Gene as database entry head(kegg_get("ece:Z5100").read())

ENTRY Z5100 CDS T00044

NAME espF

DEFINITION (GenBank) espF

ORTHOLOGY K12786 LEE-encoded effector EspF

ORGANISM ece Escherichia coli O157:H7 EDL933 (EHEC)

PATHWAY ece05130 Pathogenic Escherichia coli infection

MODULE ece_M00542 EHEC/EPEC pathogenicity signature, T3SS and effectors

BRITE KEGG Orthology (KO) [BR:ece00001] Human Diseases

Retrieving gene sequences

Gene as amino acid sequence

print(kegg_get("ece:Z5100", "aaseq").read())

>ece:Z5100 K12786 LEE-encoded effector EspF | (GenBank) espF; espF (A) MLNGISNAASTLGRQLVGIASRVSSAGGTGFSVAPQA VR-LTPVKVHSPFSPGSSNVNART IFNVSSQ-VTSFTPSRPAPPPPT-SGQASGASRPLPPIAQALKEHLAAYE-KSKGPEALGFKP ARQAP-PPPTSGQASGASRPLPPIAQALKEHLAAYEK-S K G P E A L G F K P A R Q A P P P P T S G Q A SGASRPLPPIAQALKEHLAAYEKSKGPEALG-FKPARQAPP-PPTGPSGLPPLA-QALKDHLA AYEQSKKG

Gene as nucleotide sequence

print(kegg_get("ece:Z5100", "ntseq").read())

>ece:Z5100 K12786 LEE-encoded effector EspF | (GenBank) espF; espF (N) atgcttaatggaattagtaacgctgcttctacactagggcggcagcttgtaggtatcgca agtcgagtgagctctgcgggggggaactggattttctgtagcccctcaggccgtgcgtctt actccggtgaaagttcattcccctttttctccaggctcgtcgaatgttaatgcgagaacg atttttaatgtgagcagccaggtgacttcatttactccctctcgtccggcaccgccgcca ccgacaagtggacaggcatccggggcatcccgacctttaccgcccattgcacaggcatta aaagagcacttggctgcctatgaaaaatcgaaaggtcctgaggctttaggttttaagccc gcccgtcaggcaccgccgccaccgacaagtggacaggcatccggggcatcccgaccttta ccgcccattgcacaggcattaaaagagcacttggctgcctatgaaaaatcgaaaggtcct gaggctttaggttttaagcccgcccgtcaggcaccgccgccaccgacaagtggacaggca tccggggcatcccgacctttaccgcccattgcacaggcattaaaagagcacttggctgcc tatgaaaaatcgaaaggtcctgaggctttaggttttaagcccgcccgtcaggcaccaccg ccaccgacagggcctagtggactaccgccccttgcacaggcattaaaagatcatttagct gcctatgagcaatcgaagaaagggtaa

Parsing a returned sequence with SeqIO, and converting format

seq = SeqIO.read(kegg_get("ece:Z5100", "ntseq"), 'fasta')

print(seq.format('stockholm'))

STOCKHOLM 1.0 #=GF SQ 1 ece:Z5100 atgcttaatg-gaattagt aacgctgcttctacactagggcggca-gcttgtaggtatcgcaagtcgagtgagctctgc ggggggaactggattttctgtagccc-ctcaggccgtgcgtcttactccggtgaaagttcatt cccctttttctccag gctcgtcgaatgttaatgcg-agaacgatttttaatgtgagcagccag-gtgacttcatttactccctctcgtccggca ccgccgccaccgacaagtggacag-gcatccggggcatcccgacctttaccgcccattgcacaggcattaaaagagca cttggctgcctatgaaaa-atcgaaaggtcctgaggctttaggttttaagcccgcccgtcaggca ccgccgccaccga caagtggacaggcatccggggcatcccgacctttaccgcccattgcac-aggcattaaaagagcacttggctgcctat gaaaaatcgaaaggtcctgaggctttaggttttaagcccg cccgtcagg-caccgccgccaccgacaagtggacaggc atccggggcatcccgacctttaccgcc-cattgcacaggcattaaaaga-gcacttggctgcctatgaaaaatcgaaag gtcctgaggctttaggttttaagcc-cgcccg-tcaggcaccaccgccaccgacagggcctagtgg actaccgcccctt gcacaggcattaaaagatcatttagctgcctatgagcaatcgaagaaagggtaa

#=GS ece:Z5100 AC ece:Z5100

#=GS ece:Z5100 DE ece:Z5100 K12786 LEE-encoded effector EspF | (GenBank) espF; espF (N) //

Pathways can be returned as database entries in plain text, or in the KGML format, or as .png images. These last two formats are especially useful for generating custom pathway renderings with Bio.Graphics. KGML.

```
# Pathway as database entry
head(kegg_get("hsa05130").read())
ENTRY       hsa05130     Pathway
```

NAME Pathogenic Escherichia coli infection -Homo sapiens (human) DESCRIPTION Enteropathogenic E. coli (EPEC) and enterohemorrhagic E. coli (EHEC) are closely related pathogenic strains of Escherichia coli. The hallmark of EPEC/EHEC infections [DS: H00278 H00277] is induction of attaching and effacing (A/E) lesions that damage intestinal epithelial cells. The capacity to form A/E lesions is encoded mainly by the locus of enterocyte effacement (LEE) Pathogenicity Island. Tir, Map, EspF, EspG are known LEE-encoded effector proteins secreted via the type III secretion system, which is also LEE-encoded, into the host cell. EPEC and EHEC Tir's link the extracellular bacterium to the cell cytoskeleton. Map and EspF are involved in mitochondrion membrane permeabilization. EspG interacts with tubulins and stimulates microtubule destabilization. LEE-encoded adhesin or intimin (Eae) is exported via the general secretory pathway to the periplasm, where it is inserted into the outer membrane. In addition to Tir, two potential host cell-carried intimin receptors, beta1 integrin (ITGB1) and nucleolin (NCL), have so far been identified. The distinguishing feature of EHEC is the elaboration of Shiga-like toxin (Stx). Stx cleaves ribosomal RNA, thereby disrupting protein synthesis and killing the intoxicated epithelial or endothelial cells.

CLASS Human Diseases; Infectious diseases

PATHWAY_MAP hsa05130 Pathogenic Escherichia coli infection

[...]

```
# Pathway as image (png)
Image(kegg_get("hsa05130", "image").read())
```

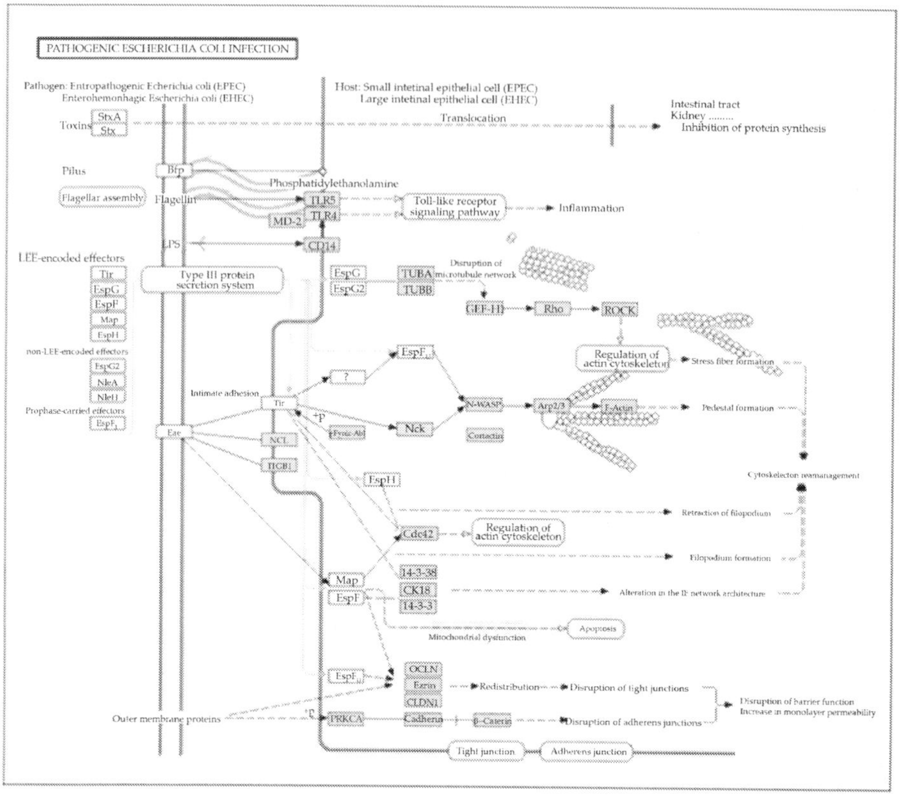

Pathway as KGML

```
head(kegg_get("hsa05130", "kgml").read())
<?xml version="1.0"?>
<!DOCTYPE pathway SYSTEM "http://www.kegg.jp/kegg/xml/KGML_v0.7.2_.dtd">
<!-- Creation date: Feb 24, 2016 14:42:27 +0900 (GMT+9) -->
<pathway name="path: hsa05130" org="hsa" number="05130" title="Pathogenic Escherichia coli infection" image="http://www.kegg.jp/kegg/pathway/hsa/hsa05130.png" link="http://www.kegg.jp/kegg-bin/show_pathway?hsa05130">
    <entry id="1" name="path:hsa04810" type="map" link="http://www.kegg.jp/dbget-bin/www_bget?hsa04810">
```

<graphics name="Regulation of actin cytoskeleton" fgcolor="#000000" bgcolor="#FFFFFF"

[...]

RENDERING PATHWAYS WITH BIO.KEGG AND BIO.GRAPHICS.KGML

The easiest way to render a pathway is to retrieve an image file directly from KEGG with *kegg_get()*, as above.

Render central metabolism

Image(kegg_get("map01100", "image").read())

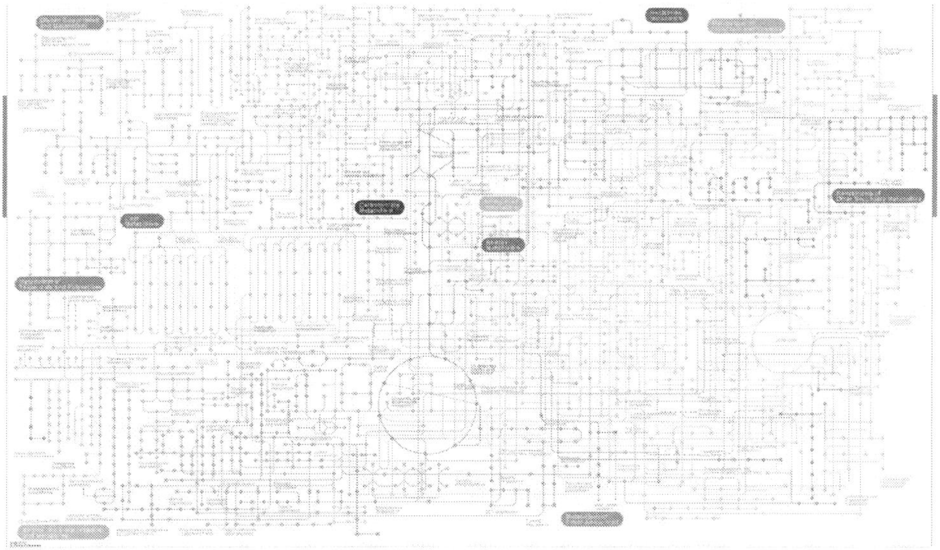

Render fatty-acid biosynthesis

Image(kegg_get("map00061", "image").read())

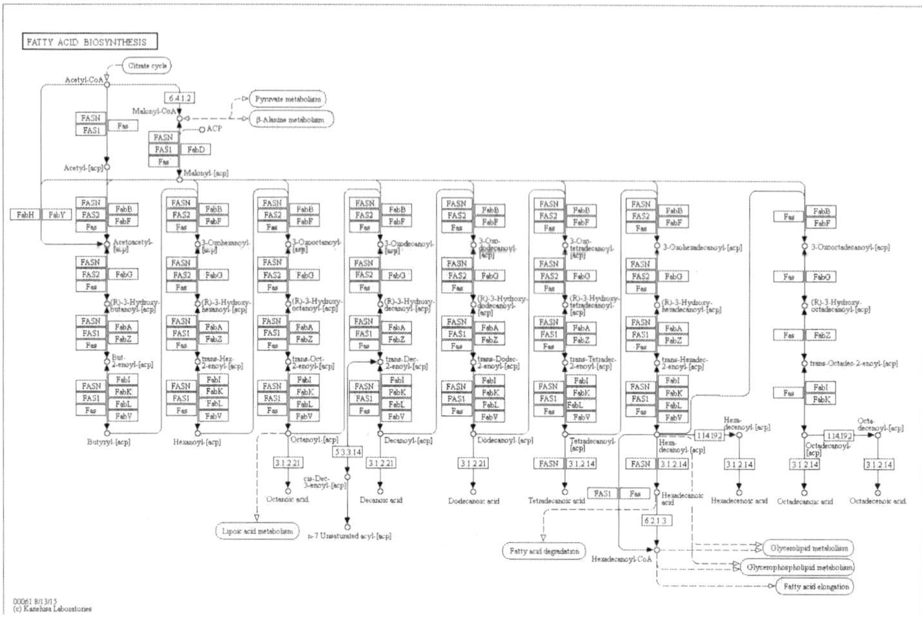

KEGG also provides renderings that are specific to an organism, indicating which elements of the pathway are present in their database for that organism. For central metabolism, compounds and reactions are only coloured if they are present in the specified organism. Otherwise, they are rendered in grey.

```
# Render E.coli K-12 MG1655 central metabolism
Image(kegg_get("eco01100", "image").read())
```

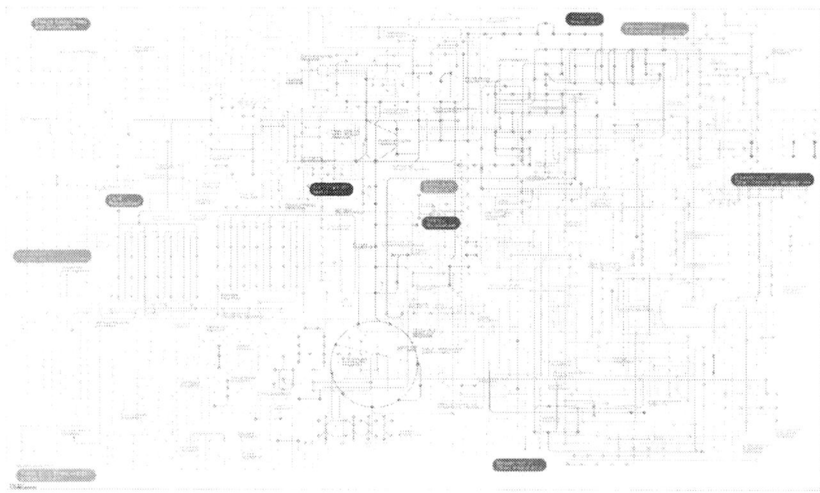

With other pathways, the reactions and/or genes are coloured green when they are present in the organism, and in white otherwise.

```
# Render E.coli K-12 MG1655 fatty-acid biosynthesis
Image(kegg_get("eco00061", "image").read())
```

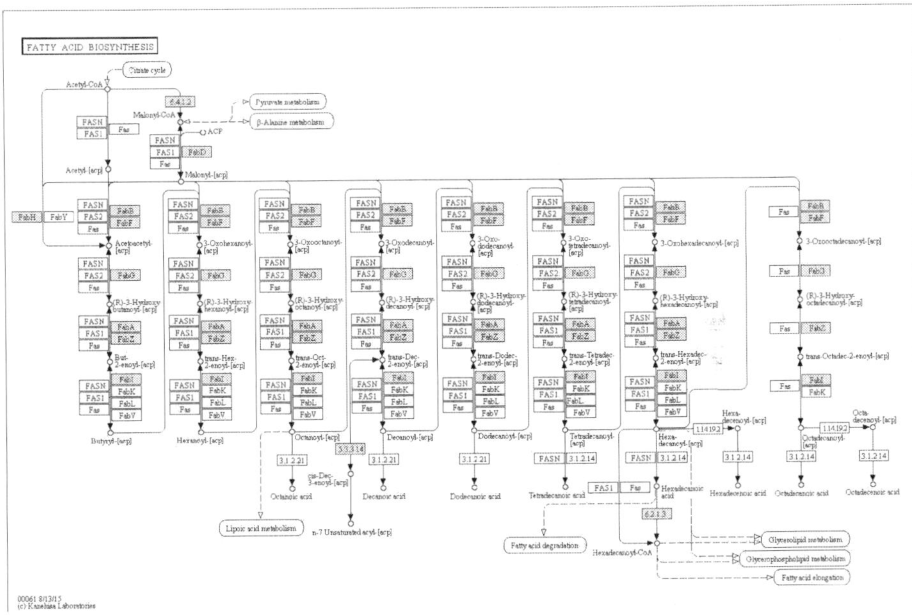

Three types of reference pathway can also be rendered: ko, ec, and rn. One of these are shown with blue boxes. The reader can try the other commented commands.

```
# Render reference fatty-acid biosynthesis
Image(kegg_get("ko00061", "image").read())
#Image(kegg_get("ec00061", "image").read())
#Image(kegg_get("rn00061", "image").read())
```

SUMMARY

We have introduced a very useful database in computational molecular biology known as KEGG. We introduce the reader to the basic organization of the database and have provided a short tutorial on accessing the data programmatically.

REFERENCES

1. Kanehisa M, Goto S (2000). "KEGG: Kyoto Encyclopedia of Genes and Genomes." *Nucleic Acids Res.* **28** (1): 27–30.
2. https://en.wikipedia.org/wiki/KEGG
3. https://github.com/widdowquinn/Notebooks-Bioinformatics/blob/f1d6f89eb629bc30a3e712139a1187e58df47119/Biopython_KGML_intro.ipynb

8

BIOMOLECULAR STRUCTURE AND CONFORMATION

We have spent quite some time understanding biomolecular sequences and their processing using computational techniques. In particular, we have learnt processing sequence data in **Python** and **R**. In the present chapter, we will shift our focus on using computation to comprehend biomolecular structures. We will be using both **R** and **Python** specialized packages for this purpose. We will focus primarily on protein structures, as understanding the biology of proteins is dependent much more on the precise three-dimensional structures. In comparison, DNA has a repetitive double helix structure and is quite inert. Protein structures present more variety. The structures can be comprehended at any level(s), from atoms to complexes with other biomolecules (proteins, DNA/RNA, peptides etc). We start with an introduction to protein structure and venture into computational manipulation of structure files.

PROTEINS AND STRUCTURES

Proteins are three dimensional objects. The sequence of amino acid residues assumes a definite shape in space. To understand this structure, one needs a physical technique which can probe the organization of residues at atomic level. Scientists employ physical techniques like x-ray crystallography, NMR (nuclear magnetic resonance) spectroscopy, cryo-electron microscopy to probe the molecular structure. We will focus on computational techniques for manipulating and analyzing the structures determined by *x-ray crystallography*.

Proteins take up globular structures, where consecutive residues fold into three dimensional shapes. This happens in the presence of water molecules, which serves as a biological solvent. The structure formation is essentially due to the weak, non-covalent interactions among protein residues. Membrane proteins and some proteins fold in a different way to assume structures in a cell.

Let us understand in simpler terms the *folding* of proteins. The atoms around the peptide links, which form the backbone of a protein's polypeptide chain, are capable of interacting in a stabilising way with water and amongst themselves; the *amide* (N-H) and *carboxyl* (C=O) groups form polar *hydrogen bonds*. All things being equal the interaction with water is stronger, but the other parts of the amino acids that stick out from their backbone, the *side chains*, tip the balance so that the protein backbone is mostly stabilised by the backbone atoms hydrogen bonding with each other, and not water. The different amino acids have chemical structures that govern whether their side chain can make a significant interaction with water. Side chains containing atomic groups that can form relatively strong hydrogen bonds (O-H, N-H, C=O) and those that carry an electric charge are said to be *hydrophilic* (water-loving), because they can make stabilising interactions with water. Residues that do not form such interactions are described as *hydrophobic* (water-hating). Strictly speaking there is not a set dividing line between hydrophobic and hydrophilic; it is more a matter of degree. In an aqueous (water) environment, the hydrophobic and hydrophilic residues segregate when a protein folds, to form a hydrophobic core and hydrophilic exterior, i.e. a globule. This is just a general trend though; the protein globule is stabilised further by the hydrogen bonds along the backbone, which tend to form regular patterns of hydrogen-bonding networks, called *secondary structure*. Also, the electric charges and polarities will push and pull the structure into the final shape. This final conformation is one where the core residues (mostly hydrophobic) come together and give rise to another weaker, but widespread, kind of interaction described as the *van der Waals* force, and thus the core packs tightly. This weak nonbonding interaction is actually present all the time between close atoms, including those from water, but in many situations it is swamped by other, stronger interactions. As a final point on protein folding, it should be noted that some large sections of amino acid sequences do not have a significant hydrophobic component. These regions will typically not form a single stable, folded structure because they don't have the ability to form a hydrophobic core. Usually this results in the region being highly dynamic or *unstructured* and is commonly seen at the ends of protein chains and as flexible linkers between folded *domains*, which are compact and globular.

DETERMINING BIOMOLECULAR STRUCTURES

X-ray crystallography, as you might guess, works by forming crystals made of the molecule of interest, and is the most common form of macromolecular structure determination. Growing the crystals to start with is often the most difficult part of the process. However, if good crystals can be made, then because the molecules are held together in a lattice, the regular array of atom positions acts to diffract a beam of X-ray radiation. Thus, creating a *diffraction pattern* consisting of spots called *reflections,* once the beam has passed though the crystal. Several diffraction patterns are collected by rotating the sample in a precisely known way, to view the structure from a variety of angles. The diffraction patterns can be combined, by mathematical means that we will not go into, to derive a three-dimensional map of the *electron density* of the atoms. The structure of the macromolecule is then determined by fitting its covalent chemical structure into the electron density to find the best fit. For proteins this usually involves a computer program that first finds the backbone route (fairly easy because of the characteristic spacing and angles along the polypeptide) and then fits the side chains, although this last part often needs manual assistance.

Nuclear magnetic resonance (NMR) is the second most common form of macromolecular structure determination. Although overall many more structures have been determined by crystallography, NMR is particularly common for smaller structures, where it works best; indeed there is generally a limit to the size of the molecule that can be studied this way.

NMR typically uses a concentrated solution of the subject (although solid-state samples are becoming more common), which is then placed in an extremely powerful electromagnet. In such a strong magnetic field some of the atomic nuclei, those that have *spin-active* isotopes, will align with the magnetic field. While the spin-active hydrogen-1 and phosphorus-31 isotopes are very abundant in nature, the carbon-13 and nitrogen-15 isotopes are rare and thus biological samples are usually artificially enriched with these. The magnetic alignments of spin-active atomic nuclei are then detected by passing a pulse of radio waves though the sample, so that they resonate. Because the resonance frequencies of the atoms depend on their chemical and structural environment, individual atoms often have distinct resonant frequencies. By the use of specially designed radio pulses the magnetisation can be moved from one atom to another, selectively to covalently bound atoms or those that are close in space. Such experiments correlate the observed resonances so that, often after a lengthy analysis, they can be connected together to identify which resonance goes with which atom and then how the resonances are arranged in space, thus yielding a three-dimensional structure.

We will not consider comparative modelling in this text, as the scope of the subject is very large and cannot be discussed at length in this book. For a greater in-depth understanding of X-ray Crystallography and the techniques involved please refer to International Tables of Crystallography Volume F.

FROM 1-DIMENSIONAL SEQUENCE TO 3-DIMENSIONAL STRUCTURE

There are four distinct levels of protein structure. The basic unit of proteins is amino acid residues. Each α-amino acid consists of a backbone that is present in all the amino acid types and a side chain that is unique to each type of residue. An exception from this rule is proline. Because, the carbon atom is bound to four different groups it is *chiral*. However, only one of the isomers occur in biological proteins. Glycine is not chiral since its side chain is a hydrogen atom. A simple mnemonic for correct L-form is "CORN": when the C^{α} atom is viewed with the H in front, the residues read "CO-R-N" in a clockwise direction.

Now, let us look at the basics of a protein structure.

PRIMARY STRUCTURE

The primary structure of a protein refers to the linear sequence of amino acids in the polypeptide chain. The primary structure is held together by covalent bonds known as *peptide bonds*, which are made during the process of protein biosynthesis. The two ends of the polypeptide chain are referred to as the carboxyl terminus (C-terminus) and the amino terminus (N-terminus). Counting of residues always starts at the N-terminal end (NH_2-group), which is the end where the amino group is not involved in a peptide bond. The primary structure of a protein is determined by the gene coding for the protein. A specific sequence of nucleotides in DNA is transcribed into mRNA, which is read by the ribosome in a process called **translation**. The sequence of amino acids in **insulin** was discovered by Frederick Sanger, establishing that proteins have *defined* amino acid sequences. The sequence of a protein is unique to that protein, and defines the structure and function of the protein.

The primary sequence of a protein can be determined by methods such as **Edman degradation** or **tandem mass spectrometry**. Often, however, it is read directly from the sequence of the gene using the genetic code. It is strictly recommended to use the words “amino acid residues” when discussing proteins because when a peptide bond is formed, a water molecule is lost, and therefore proteins are made up of amino acid residues. Post-translational modification (PTM) such as disulfide bond formation, phosphorylations and glycosylations are also considered as part of the primary structure, and cannot be read from the gene. For example, insulin is composed of 51 amino acids in 2 chains. One chain has 31 amino acids, and the other has 20 amino acids. Computationally we represent the 1D sequence in the form a FASTA sequence.

SECONDARY STRUCTURE

Secondary structure refers to highly regular local sub-structures on the actual polypeptide backbone chain. Two main types of secondary structure, the a-helix and the β-sheets, were suggested in 1951 by Linus Pauling and coworkers. These secondary structures are defined by patterns of hydrogen bonds between the main-chain peptide groups. They have a regular geometry, being constrained to specific values of the dihedral angles ψ and φ on the **Ramachandran** plot. Both the a-helix and the β-sheet represent a way of saturating all the hydrogen bond donors and acceptors in the peptide backbone. Some parts of the protein are ordered but do not form any regular structures. They should not be confused with random coil, an unfolded polypeptide chain lacking any fixed three-dimensional structure. Several sequential secondary structures may form a "*supersecondary unit*".

Ramachandran plot is a short-hand of representing protein secondary structural geometry. It is also called as dihedral plot, as it shows a scatter of backbone dihedral angles of each amino acid residue in a protein structure. A Ramachandran plot showing the most common secondary structures is shown below. One can see that the most commonly observed secondary structures in proteins are found to be populated on the left side of the plot. Left-handed helix is shown to the right of the Ramachandran map.

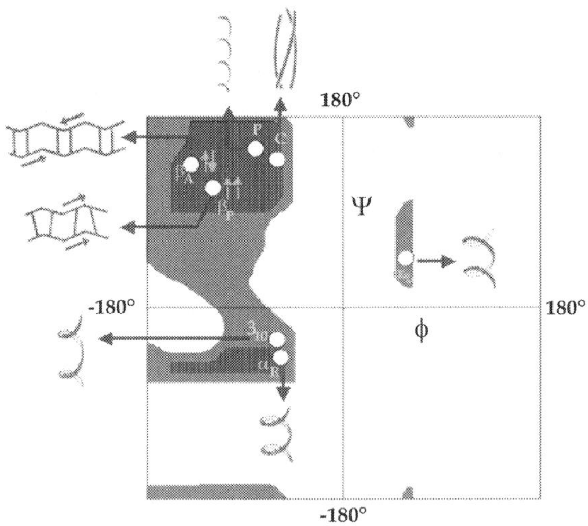

Figure 8.1: Ramachandran plot highlighting the regular secondary structural elements of proteins. α_R - right-handed alpha-helix, a_L - left-handed alpha-helix, β_P - parallel sheet, β_A - anti-parallel sheet, P - polyProline, C - collagen triple helix.

Helices

Helices are the most conspicuous elements of the protein secondary structure. Linus Pauling's prediction of the a-helix, one of the greatest achievements in structural biology, was made by assuming (i) that the peptide bond is planar, (ii) that all amino acid residues are equivalent with respect to the backbone conformations and (iii) that each amide proton is hydrogen bonded to an oxygen atom of another residue with an N-O distance of 2.72 Å. Similar criteria allowed Pauling to predict several other protein secondary structural elements, including the γ- helix and parallel and anti-parallel β-sheets. Subsequent single crystal X-ray diffraction studies revealed a-helices in the proteins hemoglobin and myoglobin. The helices found in the protein structures normally belong to the two broad classes: a-helix and 3_{10}-helix. These are characterized when a stretch of consecutive residues adopt nearly identical (f, y) values, given planar, *trans*-peptide amide bonds. The high resolution crystal structures of synthetic peptides have revealed the structural aspects of helical conformations and have provided precise geometric and conformational parameters that characterize the wide variety of helices. The a-helix is stabilized by hydrogen bonds between the CO group of residue (*i*) to the NH group of residue (*i*+4) (5→1 hydrogen bond).

The second most frequently observed helical structure in proteins is the 3_{10}-helix. The 3_{10}-helix is characterized by a hydrogen bond between the CO group of residue (*i*) and the NH group of residue (*i*+3) (4→1 hydrogen bond) with 10 atoms in the hydrogen bonded ring. The 3_{10} helix is found to occur at the terminal regions of a-helices and in short loops.

Helices with 6→1 hydrogen bonds (π-helix or 4.2_{16}-helix) have also been characterized in crystal structures of proteins, although they are less frequently observed. Figure below shows a view of ideal a-helix, 3_{10}-helix and π-helix.

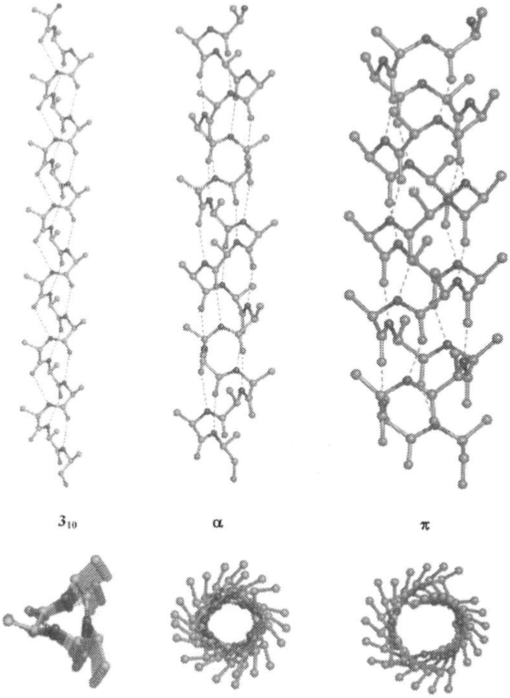

Figure 8.2: Three dimensional models of ideal 3_{10}-, a-, and π-helices. Figures below are projections down the helical axis.

Sheets and Hairpins

β-sheet is the next most important secondary structure of proteins. In contrast to the a-helix, which is built from one continuous region, a β-sheet is built up from a combination of several regions of the polypeptide chain. Pauling and Corey were first to model the sheet and David Philips and colleagues were first to see the sheet in the structure of lysozyme. The (f, y) values adopted by β-sheets fall in the broad structurally allowed region in the upper left quadrant of the Ramachandran plot. A β-sheet is built by the β-strands. The β-strands of the sheets are aligned adjacent to each other and have a "pleated" appearance, with the C^a atoms successively a little above and below the plane of the β-sheet. β-strands can interact in two ways to form pleated β-sheet; parallel (amino acids in the aligned strands run in the same direction, amino terminal to carboxy terminal) and anti-parallel (amino acids in the aligned strands run in the opposite directions, amino terminal to carboxy terminal then carboxy terminal to amino terminal and soon). β-strands can combine into mixed β-sheets with some β-strand pairs parallel and some anti-parallel. The pattern of hydrogen bonding between the amide and carbonyl groups of the protein backbone is distinctive of β-sheets. In parallel β-sheets, where the β-strands run in the same amide-to-carbonyl direction, the backbone hydrogen bonds are evenly spaced and angle across to the adjacent main chain. In contrast, the hydrogen bonds formed in anti-parallel sheets are approximately perpendicular to the main chain. β-sheet side chains can alternate above and below the plane of the sheet along each strand. On any one face, the direction of two adjacent "cross-strand" side chains is in register. In anti-parallel β-sheets, there are two types of symmetrical cross-strand pairs: a narrow hydrogen-bond pair and a wide pair, also referred to as H-bonded and non-H-bonded pairs. The β-strands found in proteins are observed to possess a right-handed twist, favoured by inter-strand non-bonded interactions and geometric constraints.

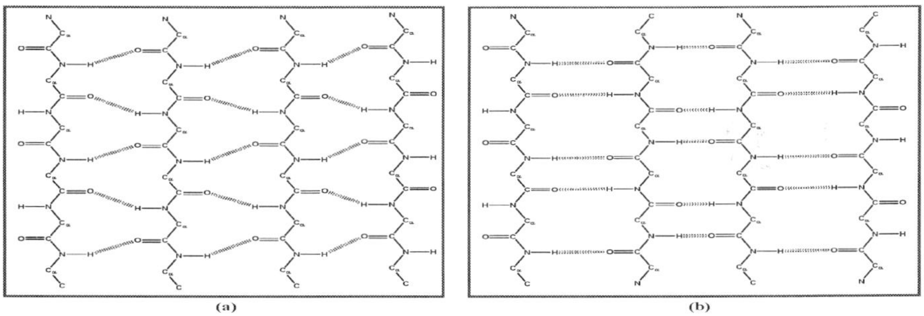

Figure 8.3: Schematic representation of (a) parallel and (b) anti-parallel β-sheets.

TERTIARY STRUCTURE

Tertiary structure refers to the three-dimensional structure of monomeric and multimeric protein molecules. The α-helixes and β-pleated-sheets are folded into a compact globular structure. The folding is driven by the *non-specific* hydrophobic interactions, the burial of hydrophobic residues from water, but the structure is stable only when the parts of a protein domain are locked into place by *specific* tertiary interactions, such as **salt bridges**, **hydrogen bonds**, and the tight packing of side chains and **disulfide bonds**. The disulfide bonds are extremely rare in cytosolic proteins, since the cytosol (intracellular fluid) is generally a reducing environment.

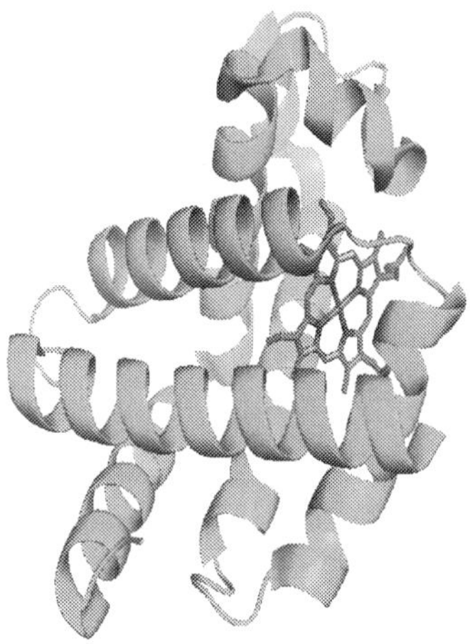

Figure 8.4: Structure of myoglobin with **heme** ligand shown as sticks (PDB: 3RGK).

QUATERNARY STRUCTURE

Quaternary structure is the three-dimensional structure of a multi-subunit protein and how the subunits fit together. In this context, the quaternary structure is stabilized by the same non-covalent interactions and disulfide bonds as the tertiary structure. Complexes of two or more polypeptides (i.e. multiple subunits) are called *multimers*. Specifically it would be called a *dimer* if it contains two subunits, a *trimer* if it contains three subunits, a *tetramer* if it contains four subunits, and a *pentamer* if it contains five subunits. The subunits are related to one another by *symmetry operations*, such as a 2-fold axis in a dimer. Multimers made up of identical subunits are referred to with a prefix of "*homo-*" (e.g. a homotetramer) and those made up of different subunits are referred to with a prefix of "*hetero-*", for example, a heterotetramer, such as the two alpha and two beta chains of hemoglobin.

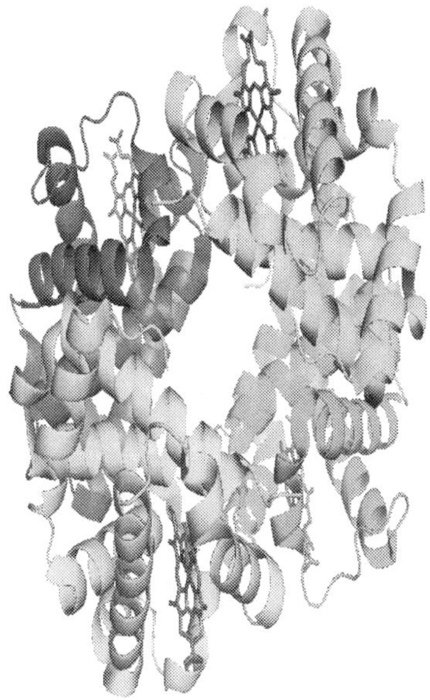

Figure 8.5: Structure of human hemoglobin (PDB: 1GZX). The four identical subunits of hemoglobin are shown in four different colors. Heme molecules are shown as sticks.

BIOMOLECULAR CONFORMATION

We have seen that the basic units of proteins are the naturally occurring 20 amino acid residues. All the 20 amino acid residues have an identical backbone. Two consecutive amino acid residues in a sequence combine by forming a peptide bond between the basic amino group and the acidic carboxyl group of each amino acid. This results in a long chain of amino acids with a repeating backbone.

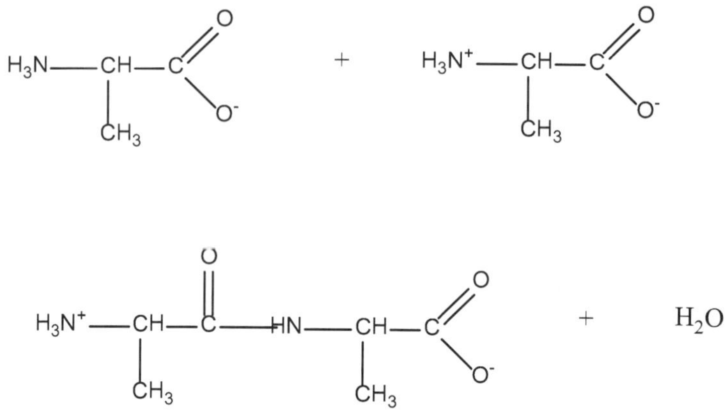

Figure 7.6: The formation of a peptide bond.

The above figure illustrates the condensation of two alanine (Ala) residues. In principle, the same process repeats for the condensation of any class of residues. The basic chemical structure of the peptide backbone was first described by Linus Pauling and colleagues. From the known structures of amino acids, Pauling and Corey first predicted the existence of alpha helices and beta sheets as a component of protein structure. Ramachandran first described exactly what range of conformations are available to amino acids in a peptide chain. Peptide chain conformation is simply described by the values of the dihedral angles in the protein backbone (i.e., the angle described by the four atoms surrounding N-C^a bond and the angle described by the four atoms surrounding the C^a-C bond). These angles are referred to as f and y, respectively. The chain isn't free to rotate around the third kind of bond in the protein backbone, the peptide bond, because it is a partial double bond and hence chemically constrained to be planar, so the values of f and y for each amino acid provide a complete description of the protein backbone. A Ramachandran map is simply a plot of f versus y for an entire protein structure (Figure 7.1). One means of evaluating a protein structure model is to compare its individual Ramachandran map with the general Ramachandran map of allowed values of (f, y).

It's apparent from the Ramachandran map that *steric* interactions are very important determinants of the general features of protein structure. Steric interactions instantly eliminate a large fraction of possible conformations for proteins and leave relatively few options for how a compact structure can form from a linear chain of amino acids. The sequence of a protein is called its primary structure; the most basic level of organization in a protein is the sequence of amino acids. Alpha helix and beta sheet structures, are known collectively as secondary structures and are the next level of organization. Interactions between multiple secondary structure elements give rise to supersecondary structure and tertiary structure—helices and sheets contacting each other to form larger characteristic structures, which can be described by their topology.

Hydrogen bonds

Hydrogen bonds arise when two *polar* groups interact. The two polar groups must be of specific types. One must be a proton *donor*, a chemical group in which a proton (hydrogen atom) is covalently bonded to a strongly *electronegative* atom such as oxygen. The bond between the proton and the electronegative atom is polarized, giving the proton a partial positive charge and the electronegative atom a partial negative charge. The other group must be a proton *acceptor*, an electronegative atom with a partial negative charge and no attached proton. The positively polarized proton in the first group is attracted to the negatively polarized second group, and the two form a bond. The bond formed isn't covalent, but much shorter and stronger than a normal non-bonded interaction. Hydrogen bonds are unusual among non-bonded and electrostatic interactions. This is because hydrogen bonds are strongly directional; they weaken if the angle described by the three atoms involved is too large or too small. Linear hydrogen bonds are the strongest.

Hydrogen bond interactions are one of the most important stabilizing forces in protein structure. The protein backbone contains a proton donor, in its N-H group, and a proton acceptor, in its carbonyl oxygen, spaced at regular intervals along the chain. The interaction of these groups stabilizes the two major types of secondary structure, the alpha helix and the beta sheet. Therefore, some structure prediction methods attempt to use the presence of potential hydrogen bond pairs to improve the accuracy of predictions.

Hydrophobic and Hydrophilic Interactions

We've already mentioned in passing that amino acids can be classified as hydrophobic or hydrophilic. What exactly does this mean? Proteins, except for those bound within cell membranes, always exist in aqueous solution. They constantly interact with water molecules. Water is a solution that has some interesting properties, and these properties contribute to the stability of the compact globular structures that characterize cellular proteins.

- Water is a polar molecule.
- Individual water molecules in liquid water can each form four hydrogen bonds with neighbouring water molecules.

Liquid water is an essentially uninterrupted lattice of hydrogen bonded molecules. This unusual property contributes to the high melting and boiling points of water, as well as to such properties as low compressibility and high surface tension. It also results in interesting interactions of water with soluble proteins.

Consider a non-polar molecule dissolved in water. This molecule interrupts the regular hydrogen bond lattice of liquid water. Individual water molecules can reorient around a small non-polar molecule to preserve their network of hydrogen bonds, but this *reorientation* has a cost in terms of *free energy* (which is how cost is measured in chemistry). The presence of a non-polar solute forces water molecules into a more ordered conformation than they would ordinarily assume. Instead of being able to face any which way and rotate freely, water molecules near the surface of a non-polar solute have to work around it and form a cage. This is entropically not favourable.

The larger a non-polar solute, the more the water molecules have to reorient to accommodate it, and the higher the energy cost of solvating the molecule becomes. Of course, if the non-polar solute has some polar groups on its surface, water molecules can use those groups as hydrogen bonding partners instead of other water molecules, and the water strcture is less disturbed. Globular proteins, which exist in aqueous solution even though they are composed substantially of non-polar groups, must present a good hydrogen-bonding surface to the aqueous environment. *Hydrophilic* amino acids are those whose side-chains offer hydrogen bonding partners to the surrounding medium, while *hydrophobic* amino acids' side-chains resist bonding with water. The surface of a globular protein is usually anywhere from 50%-75% polar atoms, and deviations in this pattern can suggest binding or complexation sites. This causes the hydrophilic molecules to get exposed on the surface and hydrophobic groups to come together to form the core of the protein.

Solvent accessibility and hydrophobicity play an important role in evaluating model structures. Threading methods for protein fold recognition use amino acid environments in evaluating models. When many hydrophobic amino acids are found in solvent-exposed structural environments or hydrophilic amino acids buried in the protein interior, it is considered unlikely that the protein model is folded correctly.

Charge-Charge, Charge-Dipole, and Dipole-Dipole Interactions

Unlike covalent bonds, the other important interactions in protein structure are non-specific. They don't change the discrete nature of the interacting atoms. They involve no sharing of electrons. Covalently bonded atoms are married; non-covalently bonded atoms are just interacting in a slightly transient way.

Several kinds of important forces can arise among polar and charged atoms. An *ion* is an atom that has a net positive or negative charge due to either a surplus or a deficit of electrons. Atoms that carry a positive ionic charge are attracted to atoms that carry a negative ionic charge, with a strength that depends on the size of the charges and the inverse of the distance between the atoms. Now in proteins, charge-charge interactions occur between the side-chains of acidic and basic amino acids that are negatively charged or positively charged due to loss or gain of a labile proton under normal physiological conditions. The charge-charge interactions between amino acids in a protein structure are called *salt bridges*, and they can contribute a significant stabilizing force to a protein structure.

There are other, weaker interactions that occur between charges and groups that don't carry a positive or negative ionic charge. *Dipolar molecules* are molecules like those involved in hydrogen bonds, in which one end of the molecule has a partial positive charge and the other end has a partial negative charge. The dipole of a molecule is essentially a vector that describes the magnitude of the polarization along a bond. Dipolar molecules can be strongly attracted to other partial charges or to ionic charges. Many amino acid side-chains, as well as the protein backbone, have a strongly dipolar character, so charge-dipole and dipole-dipole interactions play a substantial role in stabilization of protein structure.

Van Der Waals Forces

The van der Waals force is a non-specific attractive force between molecules. This force is loosely analogous to gravity, in that it exists between every pair of non-bonded atoms, and it's a fairly long-range force. However, it doesn't arise simply from the mass of the atoms involved, but from the transient attractive forces between the instantaneous dipole moments of each atom. The van der Waals force is quite strong, and because van der Waals interactions are non-specific and numerous they play a significant role in protein folding and protein association.

Repulsive Forces

Repulsive forces, or *steric* interactions, are very short range forces that increase sharply as atomic centers approach each other. The radius at which the repulsive force begins to increase sharply defines a spherical boundary around each atom center inside which another atom's spherical boundary (called the van der Waals radius) can't pass. If two non-bonded atoms in a structure get into each other's personal space, the contact is energetically unfavorable. In real molecules, atoms stay out of each other's way. However, in models of molecules, whether derived from NMR or x-ray data or built from scratch, checking for van der Waals bumps between non-bonded atoms is an important part of the structure-refinement process.

There are a lot of theoretical concepts in the subject of biomolecular structures. We have covered some of the primary concepts in understanding structure of biomolecules. We shift our focus to processing the structural information using computational techniques. We will use BioPython's Bio.PDB module for doing such manipulations.

PYTHON FOR MACROMOLECULAR STRUCTURES

We will be using Python in this section to probe macromolecular structures. we will mostly examine and manipulate existing structural data, i.e. the coordinates of atoms. The idea is that you should become familiar with how to handle structural information. We deliberately avoid going into the computational aspects of how to determine structures in the first place. We will leave such vast and specialist topics to your future diligence.

Obtaining Structural Information

Before we can begin to manipulate macromolecular structure data we must initially get hold of the coordinate information. We use the power of Python to download data directly from the PDB websites download service. The following code achieves this by making use of the *urllib* module (in Python 3; *urllib2* in Python 2), which is a standard part of a Python installation. This module will do all the hard work and we will use it to send a request to the PDB web service, the response to which will be a plain text file containing the required structural data, and all we need to do is specify the identifier (pdbId) of the entry that we wish to download.

We will be using well-tested functions written in **BioPython** library for this purpose.

Bio.PDB is a Biopython module that focuses on working with crystal structures of biological macromolecules. Among other things, Bio.PDB includes a PDBParser class that produces a Structure object, which can be used to access the atomic data in the file in a convenient manner. There is limited support for parsing the information contained in the PDB header.

Reading A Pdb File

First we create a PDBParser object:

```
>>> from Bio.PDB.PDBParser import PDBParser
```

```
>>> p = PDBParser(PERMISSIVE=1)
```

The PERMISSIVE flag indicates that a number of common problems associated with PDB files will be ignored (but note that some atoms and/or residues will be missing). If the flag is not present a PDBConstructionException will be generated if any problems are detected during the parse operation.

The Structure object is then produced by letting the PDBParser object parse a PDB file (the PDB file in this case is called 'pdb1fat.ent', '1fat' is a user defined name for the structure):

```
>>> structure_id = "1fat"
>>> filename = "pdb1fat.ent"
>>> s = p.get_structure(structure_id, filename)
```

You can extract the header and trailer (simple lists of strings) of the PDB file from the PDBParser object with the get_header and get_trailer methods. Note however that many PDB files contain headers with incomplete or erroneous information. Many of the errors have been fixed in the equivalent mmCIF files. *Hence, if you are interested in the header information, it is a good idea to extract information from mmCIF files using the MMCIF2Dict tool described below, instead of parsing the PDB header.*

Now that is clarified, let's return to parsing the PDB header. The structure object has an attribute called header which is a Python dictionary that maps header records to their values.

Example:

```
>>> resolution = structure.header['resolution']
>>> keywords = structure.header['keywords']
```

The available keys are name, head, deposition_date, release_date, structure_method, resolution, structure_reference (which maps to a list of references), journal_reference, author, and compound (which maps to a dictionary with various information about the crystallized compound).

The dictionary can also be created without creating a Structure object, i.e., directly from the PDB file:

```
>>> file = open(filename, 'r')
```

```
>>> header_dict = parse_pdb_header(file)
>>> file.close()
```

Reading Mmcif File

Similarly to the case the case of PDB files, first create an MMCIFParser object:

```
>>> from Bio.PDB.MMCIFParser import MMCIFParser
>>> parser = MMCIFParser()
```

Then use this parser to create a structure object from the mmCIF file:

```
>>> structure = parser.get_structure('1fat', '1fat.cif')
```

To have some more low level access to an mmCIF file, you can use the MMCIF2Dict class to create a Python dictionary that maps all mmCIF tags in an mmCIF file to their values. If there are multiple values (like in the case of tag _atom_site.Cartn_y, which holds the *y* coordinates of all atoms), the tag is mapped to a list of values. The dictionary is created from the mmCIF file as follows:

```
>>> from Bio.PDB.MMCIF2Dict import MMCIF2Dict
>>> mmcif_dict = MMCIF2Dict('1FAT.cif')
```

Example: get the solvent content from an mmCIF file:

```
>>> sc = mmcif_dict['_exptl_crystal.density_percent_sol']
```

Example: get the list of the *y* coordinates of all atoms

```
>>> y_list = mmcif_dict['_atom_site.Cartn_y']
```

Writing A Pdb File

Use the PDBIO class for this. It's easy to write out specific parts of a structure too, of course.

Example: saving a structure

```
>>> io = PDBIO()
>>> io.set_structure(s)
>>> io.save('out.pdb')
```

If you want to write out a part of the structure, make use of the Select class (also in PDBIO). Select has four methods:

- accept_model(model)
- accept_chain(chain)
- accept_residue(residue)
- accept_atom(atom)

By default, every method returns 1 (which means the model/chain/residue/atom is included in the output). By subclassing Select and returning 0 when appropriate you can exclude models, chains, etc. from the output. Cumbersome maybe, but very powerful. The following code only writes out glycine residues:

```
>>> class GlySelect(Select):
...    def accept_residue(self, residue):
...        if residue.get_name()=='GLY':
...            return True
...        else:
...            return False
...
>>> io = PDBIO()
>>> io.set_structure(s)
>>> io.save('gly_only.pdb', GlySelect())
```

If this is all too complicated for you, the Dice module contains a handy extract function that writes out all residues in a chain between a start and end residue.

Structure Representation in Bio.Pdb Module

The overall layout of a Structure object follows the so-called SMCRA (Structure/Model/Chain/Residue/Atom) architecture:

- A structure consists of models
- A model consists of chains
- A chain consists of residues
- A residue consists of atoms

This is the way many structural biologists/bioinformaticians think about structure, and provides a simple but efficient way to deal with structure. Additional stuff is essentially added when needed. A UML diagram of the Structure object (forget about the Disordered classes for now) is shown in Fig. Such a data structure is not necessarily best suited for the representation of the macromolecular content of a structure, but it is absolutely necessary for a good interpretation of the data present in a file that describes the structure (typically a PDB or MMCIF file). If this hierarchy cannot represent the contents of a structure file, it is fairly certain that the file contains an error or at least does not describe the structure unambiguously. If a SMCRA data structure cannot be generated, there is reason to suspect a problem. Parsing a PDB file can thus be used to detect likely problems.

Structure, Model, Chain and Residue are all subclasses of the Entity base class. The Atom class only (partly) implements the Entity interface (because an Atom does not have children).

For each Entity subclass, you can extract a child by using a unique id for that child as a key (e.g. you can extract an Atom object from a Residue object by using an atom name string as a key, you can extract a Chain object from a Model object by using its chain identifier as a key).

Disordered atoms and residues are represented by DisorderedAtom and DisorderedResidue classes, which are both subclasses of the DisorderedEntityWrapper base class. They hide the complexity associated with disorder and behave exactly as Atom and Residue objects.

A flow diagram which explains the above description is shown below.

In general, a child Entity object (i.e. Atom, Residue, Chain, Model) can be extracted from its parent (i.e. Residue, Chain, Model, Structure, respectively) by using an id as a key.

```
>>> child_entity = parent_entity[child_id]
```

You can also get a list of all child Entities of a parent Entity object. Note that this list is sorted in a specific way (e.g. according to chain identifier for Chain objects in a Model object).

```
>>> child_list = parent_entity.get_list()
```

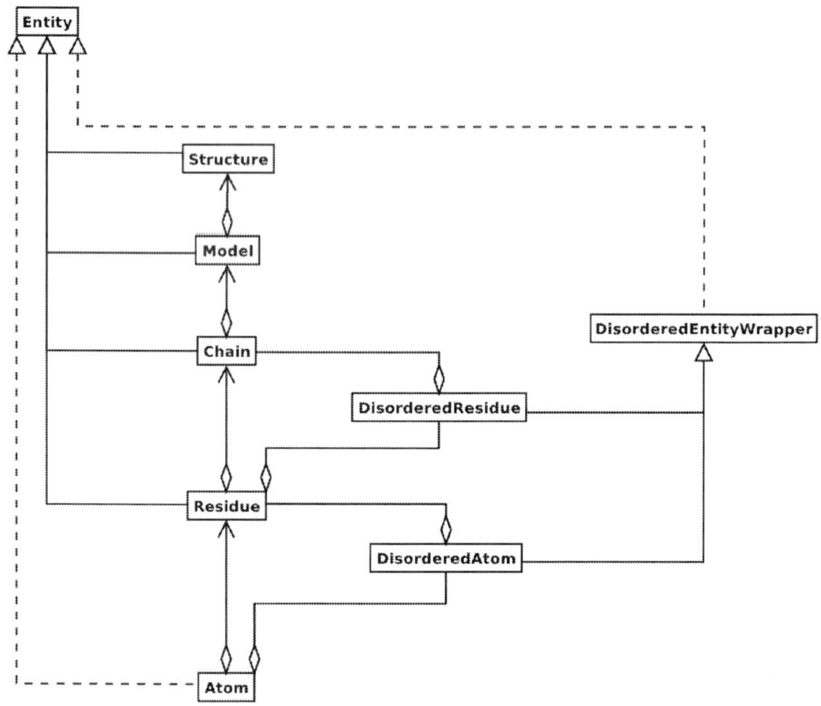

Figure 8.6: UML diagram of SMCRA architecture of the Structure class used to represent a macromolecular structure. Full lines with diamonds denote aggregation, full lines with arrows denote referencing, full lines with triangles denote inheritance and dashed lines with triangles denote interface realization.

You can also get the parent from a child:

```
>>> parent_entity = child_entity.get_parent()
```

At all levels of the SMCRA hierarchy, you can also extract a *full id*. The full id is a tuple containing all id's starting from the top object (Structure) down to the current object. A full id for a Residue object e.g. is something like:

```
>>> full_id = residue.get_full_id()
>>> print(full_id)
("1abc", 0, "A", ("", 10, "A"))
```

This corresponds to:

- The Structure with id "1abc"
- The Model with id 0

- The Chain with id "A"
- The Residue with id (" ", 10, "A").

The Residue id indicates that the residue is not a hetero-residue (nor a water) because it has a blank hetero field, that its sequence identifier is 10 and that its insertion code is "A".

To get the entity's id, use the get_id method:

```
>>> entity.get_id()
```

You can check if the entity has a child with a given id by using the has_id method:

```
>>> entity.has_id(entity_id)
```

The length of an entity is equal to its number of children:

```
>>> nr_children = len(entity)
```

It is possible to delete, rename, add, etc. child entities from a parent entity, but this does not include any sanity checks (e.g. it is possible to add two residues with the same id to one chain). This really should be done via a nice Decorator class that includes integrity checking, but you can take a look at the code (Entity.py) if you want to use the raw interface.

Structure Object

The Structure object is at the top of the hierarchy. Its id is a user given string. The Structure contains a number of Model children. Most crystal structures (but not all) contain a single model, while NMR structures typically consist of several models. Disorder in crystal structures of large parts of molecules can also result in several models.

Model Object

The id of the Model object is an integer, which is derived from the position of the model in the parsed file (they are automatically numbered starting from 0). Crystal structures generally have only one model (with id 0), while NMR files usually have several models. Whereas many PDB parsers assume that there is only one model, the Structure class in Bio. PDB is designed such that it can easily handle PDB files with more than one model.

As an example, to get the first model from a Structure object, use

```
>>> first_model = structure[0]
```

The Model object stores a list of Chain children.

Chain Object

The id of a Chain object is derived from the chain identifier in the PDB/mmCIF file, and is a single character (typically a letter). Each Chain in a Model object has a unique id. As an example, to get the Chain object with identifier "A" from a Model object, use

```
>>> chain_A = model["A"]
```

The Chain object stores a list of Residue children.

Residue Object

A residue id is a *tuple* with three elements:

- The **hetero-field** (hetfield): this is
    - 'W' in the case of a water molecule;
    - 'H_' followed by the residue name for other hetero residues (e.g. 'H_GLC' in the case of a glucose molecule);
    - blank for standard amino and nucleic acids.

This scheme is adopted for reasons above.

- The **sequence identifier** (resseq), an integer describing the position of the residue in the chain (e.g., 100);
- The **insertion code** (icode); a string, e.g. 'A'. The insertion code is sometimes used to preserve a certain desirable residue numbering scheme. A Ser 80 insertion mutant (inserted e.g. between a Thr 80 and an Asn 81 residue) could e.g. have sequence identifiers and insertion codes as follows: Thr 80 A, Ser 80 B, Asn 81. In this way the residue numbering scheme stays in tune with that of the wild type structure.

The id of the above glucose residue would thus be ('H_GLC', 100, 'A'). If the hetero-flag and insertion code are blank, the sequence identifier alone can be used:

```
# Full id
>>> residue=chain[(' ', 100, ' ')]
# Shortcut id
>>> residue=chain[100]
```

The reason for the hetero-flag is that many, many PDB files use the same sequence identifier for an amino acid and a hetero-residue or a water, which would create obvious problems if the hetero-flag was not used.

Unsurprisingly, a Residue object stores a set of Atom children. It also contains a string that specifies the residue name (e.g. "ASN") and the segment identifier of the residue.

Let's look at some examples. Asn 10 with a blank insertion code would have residue id (' ', 10, ' '). Water 10 would have residue id ('W', 10, ' '). A glucose molecule (a hetero residue with residue name GLC) with sequence identifier 10 would have residue id ('H_GLC', 10, ' '). In this way, the three residues (with the same insertion code and sequence identifier) can be part of the same chain because their residue id's are distinct.

In most cases, the hetflag and insertion code fields will be blank, e.g. (' ', 10, ' '). In these cases, the sequence identifier can be used as a shortcut for the full id:

```
# use full id
>>> res10 = chain[(' ', 10, ' ')]
# use shortcut
>>> res10 = chain[10]
```

Each Residue object in a Chain object should have a unique id. However, disordered residues are dealt with in a special way, as described in section 11.3.3.

A Residue object has a number of additional methods:

```
>>> residue.get_resname()      # returns the residue name, e.g. "ASN"
>>> residue.is_disordered()     # returns 1 if the residue has disordered
atoms
>>> residue.get_segid()       # returns the SEGID, e.g. "CHN1"
```

```
>>> residue.has_id(name)      # test if a residue has a certain atom
```

You can use is_aa(residue) to test if a Residue object is an amino acid.

Atom Object

The Atom object stores the data associated with an atom, and has no children. The id of an atom is its atom name (e.g. "OG" for the side chain oxygen of a Ser residue). An Atom id needs to be unique in a Residue. Again, an exception is made for disordered atoms, as described in section 11.3.2.

The atom id is simply the atom name (e.g., 'CA'). In practice, the atom name is created by stripping all spaces from the atom name in the PDB file. However, in PDB files, a space can be part of an atom name. Often, calcium atoms are called 'CA..' in order to distinguish them from Cα atoms (which are called '.CA.'). In cases were stripping the spaces would create problems (i.e., two atoms called 'CA' in the same residue) the spaces are kept.

In a PDB file, an atom name consists of 4 chars, typically with leading and trailing spaces. Often these spaces can be removed for ease of use (e.g. an amino acid C α atom is labeled ".CA." in a PDB file, where the dots represent spaces). To generate an atom name (and thus an atom id) the spaces are removed, unless this would result in a name collision in a Residue (i.e. two Atom objects with the same atom name and id). In the latter case, the atom name including spaces is tried. This situation can e.g. happen when one residue contains atoms with names ".CA." and "CA..", although this is not very likely.

The atomic data stored includes the atom name, the atomic coordinates (including standard deviation if present), the B factor (including anisotropic B factors and standard deviation if present), the altloc specifier and the full atom name including spaces. Less used items like the atom element number or the atomic charge sometimes specified in a PDB file are not stored.

To manipulate the atomic coordinates, use the transform method of the Atom object. Use the set_coord method to specify the atomic coordinates directly.

An Atom object has the following additional methods:

```
>>> a.get_name()      # atom name (spaces stripped, e.g. "CA")
>>> a.get_id()        # id (equals atom name)
>>> a.get_coord()     # atomic coordinates
>>> a.get_vector()    # atomic coordinates as Vector object
>>> a.get_bfactor()   # isotropic B factor
>>> a.get_occupancy() # occupancy
>>> a.get_altloc()    # alternative location specifier
>>> a.get_sigatm()    # standard deviation of atomic parameters
>>> a.get_siguij()    # standard deviation of anisotropic B factor
>>> a.get_anisou()    # anisotropic B factor
>>> a.get_fullname()  # atom name (with spaces, e.g. ".CA.")
```

To represent the atom coordinates, siguij, anisotropic B factor and sigatm Numpy arrays are used.

The get_vector method returns a Vector object representation of the coordinates of the Atom object, allowing you to do vector operations on atomic coordinates. Vector implements the full set of 3D vector operations, matrix multiplication (left and right) and some advanced rotation-related operations as well.

As an example of the capabilities of Bio.PDB's Vector module, suppose that you would like to find the position of a Gly residue's C^{β} atom, if it had one. Rotating the N atom of the Gly residue along the C^{α}-C bond over -120 degrees roughly puts it in the position of a virtual C^{β} atom. Here's how to do it, making use of the rotaxis method (which can be used to construct a rotation around a certain axis) of the Vector module:

```
# get atom coordinates as vectors
>>> n = residue['N'].get_vector()
>>> c = residue['C'].get_vector()
>>> ca = residue['CA'].get_vector()
# center at origin
>>> n = n - ca
```

```
>>> c = c - ca
# find rotation matrix that rotates n
# -120 degrees along the ca-c vector
>>> rot = rotaxis(-pi * 120.0/180.0, c)
# apply rotation to ca-n vector
>>> cb_at_origin = n.left_multiply(rot)
# put on top of ca atom
>>> cb = cb_at_origin+ca
```

This example shows that it's possible to do some quite nontrivial vector operations on atomic data, which can be quite useful. In addition to all the usual vector operations (cross (use **), and dot (use *) product, angle, norm, etc.) and the above mentioned rotaxis function, the Vector module also has methods to rotate (rotmat) or reflect (refmat) one vector on top of another.

Extracting a Specific Atom/Residue/Chain/Model from a Structure

These are some examples:

```
>>> model = structure[0]
>>> chain = model['A']
>>> residue = chain[100]
>>> atom = residue['CA']
```

Note that you can use a shortcut:

```
>>> atom = structure[0]['A'][100]['CA']
```

DISORDER

Bio.PDB can handle both disordered atoms and point mutations (i.e. a Gly and an Ala residue in the same position).

General Approach

Disorder should be dealt with from two points of view: the atom and the residue points of view. In general, we have tried to encapsulate all the complexity that arises from disorder. If you just want to loop over all Cα atoms, you do not care that some residues have a disordered side chain. On the other hand it should also be possible to represent disorder completely in the data structure. Therefore, disordered atoms or residues are stored in special objects that behave as if there is no disorder. This is done by only representing a subset of the disordered atoms or residues. Which subset is picked (e.g. which of the two disordered OG side chain atom positions of a Ser residue is used) can be specified by the user.

Disordered Atoms

Disordered atoms are represented by ordinary Atom objects, but all Atom objects that represent the same physical atom are stored in a DisorderedAtom object. Each Atom object in a DisorderedAtom object can be uniquely indexed using its altloc specifier. The DisorderedAtom object forwards all uncaught method calls to the selected Atom object, by default the one that represents the atom with the highest occupancy. The user can of course change the selected Atom object, making use of its altloc specifier. In this way atom disorder is represented correctly without much additional complexity. In other words, if you are not interested in atom disorder, you will not be bothered by it.

Each disordered atom has a characteristic altloc identifier. You can specify that a DisorderedAtom object should behave like the Atom object associated with a specific altloc identifier:

```
>>> atom.disordered_select('A') # select altloc A atom
>>> print(atom.get_altloc())
"A"
>>> atom.disordered_select('B') # select altloc B atom
>>> print(atom.get_altloc())
"B"
```

Disordered Residues

Common case

The most common case is a residue that contains one or more disordered atoms. This is evidently solved by using DisorderedAtom objects to represent the disordered atoms, and storing the DisorderedAtom object in a Residue object just like ordinary Atom objects. The DisorderedAtom will behave exactly like an ordinary atom (in fact the atom with the highest occupancy) by forwarding all uncaught method calls to one of the Atom objects (the selected Atom object) it contains.

Point mutations

A special case arises when disorder is due to a point mutation, i.e. when two or more point mutants of a polypeptide are present in the crystal. An example of this can be found in PDB structure 1EN2.

Since these residues belong to a different residue type (e.g. let's say Ser 60 and Cys 60) they should not be stored in a single Residue object as in the common case. In this case, each residue is represented by one Residue object, and both Residue objects are stored in a single DisorderedResidue object.

The DisorderedResidue object forwards all uncaught methods to the selected Residue object (by default the last Residue object added), and thus behaves like an ordinary residue. Each Residue object in a DisorderedResidue object can be uniquely identified by its residue name. In the above example, residue Ser 60 would have id "SER" in the DisorderedResidue object, while residue Cys 60 would have id "CYS." The user can select the active Residue object in a DisorderedResidue object via this id.

Example: suppose that a chain has a point mutation at position 10, consisting of a Ser and a Cys residue. Make sure that residue 10 of this chain behaves as the Cys residue.

```
>>> residue = chain[10]
>>> residue.disordered_select('CYS')
```

In addition, you can get a list of all Atom objects (i.e., all DisorderedAtom objects are 'unpacked' to their individual Atom objects) using the get_unpacked_list method of a (Disordered)Residue object.

Hetero residues

Associated problems

A common problem with hetero residues is that several hetero and non-hetero residues present in the same chain share the same sequence identifier (and insertion code). Therefore, to generate a unique id for each hetero residue, waters and other hetero residues are treated in a different way.

Remember that Residue object have the tuple (hetfield, resseq, icode) as id. The hetfield is blank (" ") for amino and nucleic acids, and a string for waters and other hetero residues. The content of the hetfield is explained below.

Water residues

The hetfield string of a water residue consists of the letter "W." So a typical residue id for a water is ("W", 1, " ").

Other hetero residues

The hetfield string for other hetero residues starts with "H_" followed by the residue name. A glucose molecule e.g. with residue name "GLC" would have hetfield "H_GLC." Its residue id could e.g. be ("H_GLC", 1, " ").

NAVIGATING THROUGH A STRUCTURE OBJECT

Parse A Pdb File, And Extract Some Model, Chain, Residue And Atom Objects

```
>>> from Bio.PDB.PDBParser import PDBParser
>>> parser = PDBParser()
```

```
>>> structure = parser.get_structure("test", "1fat.pdb")
>>> model = structure[0]
>>> chain = model["A"]
>>> residue = chain[1]
>>> atom = residue["CA"]
```

Iterating Through All Atoms of A Structure

```
>>> p = PDBParser()
>>> structure = p.get_structure('X', 'pdb1fat.ent')
>>> for model in structure:
...     for chain in model:
...         for residue in chain:
...             for atom in residue:
...                 print(atom)
...
```

There is a shortcut if you want to iterate over all atoms in a structure:

```
>>> atoms = structure.get_atoms()
>>> for atom in atoms:
...     print(atom)
...
```

Similarly, to iterate over all atoms in a chain, use

```
>>> atoms = chain.get_atoms()
>>> for atom in atoms:
...     print(atom)
...
```

Iterating Over All Residues of A Model

or if you want to iterate over all residues in a model:

```
>>> residues = model.get_residues()
>>> for residue in residues:
```

```
...     print(residue)
...
```

You can also use the Selection.unfold_entities function to get all residues from a structure:

```
>>> res_list = Selection.unfold_entities(structure, 'R')
```

or to get all atoms from a chain:

```
>>> atom_list = Selection.unfold_entities(chain, 'A')
```

Obviously, A=atom, R=residue, C=chain, M=model, S=structure. You can use this to go up in the hierarchy, e.g. to get a list of (unique) Residue or Chain parents from a list of Atoms:

```
>>> residue_list = Selection.unfold_entities(atom_list, 'R')
>>> chain_list = Selection.unfold_entities(atom_list, 'C')
```

For more info, see the API documentation.

Extract A Hetero Residue From A Chain (E.G. A Glucose (Glc) Moiety With Resseq 10)

```
>>> residue_id = ("H_GLC", 10, " ")
>>> residue = chain[residue_id]
```

Print All Hetero Residues In Chain

```
>>> for residue in chain.get_list():
...     residue_id = residue.get_id()
...     hetfield = residue_id[0]
...     if hetfield[0]=="H":
...         print(residue_id)
...
```

Print Out the Coordinates Of All Ca Atoms In A Structure With B Factor Greater Than 50

```
>>> for model in structure.get_list():
```

```
...     for chain in model.get_list():
...         for residue in chain.get_list():
...             if residue.has_id("CA"):
...                 ca = residue["CA"]
...                 if ca.get_bfactor() > 50.0:
...                     print(ca.get_coord())
...
```

Print Out All The Residues That Contain Disordered Atoms

```
>>> for model in structure.get_list():
...     for chain in model.get_list():
...         for residue in chain.get_list():
...             if residue.is_disordered():
...                 resseq = residue.get_id()[1]
...                 resname = residue.get_resname()
...                 model_id = model.get_id()
...                 chain_id = chain.get_id()
...                 print(model_id, chain_id, resname, resseq)
...
```

Loop Over All Disordered Atoms, And Select All Atoms With Altloc A (If Present)

This will make sure that the SMCRA data structure will behave as if only the atoms with altloc A are present.

```
>>> for model in structure.get_list():
...     for chain in model.get_list():
...         for residue in chain.get_list():
...             if residue.is_disordered():
...                 for atom in residue.get_list():
```

```
...             if atom.is_disordered():
...                 if atom.disordered_has_id("A"):
...                     atom.disordered_select("A")
...
```

Extracting Polypeptides from A Structure Object

To extract polypeptides from a structure, construct a list of Polypeptide objects from a Structure object using PolypeptideBuilder as follows:

```
>>> model_nr = 1
>>> polypeptide_list = build_peptides(structure, model_nr)
>>> for polypeptide in polypeptide_list:
...     print(polypeptide)
...
```

A Polypeptide object is simply a UserList of Residue objects, and is always created from a single Model (in this case model 1). You can use the resulting Polypeptide object to get the sequence as a Seq object or to get a list of Cα atoms as well. Polypeptides can be built using a C-N or a Cα-Cα distance criterion.

Example:

```
# Using C-N
>>> ppb=PPBuilder()
>>> for pp in ppb.build_peptides(structure):
...     print(pp.get_sequence())
...
# Using CA-CA
>>> ppb=CaPPBuilder()
>>> for pp in ppb.build_peptides(structure):
...     print(pp.get_sequence())
...
```

Note that in the above case only model 0 of the structure is considered by PolypeptideBuilder. However, it is possible to use PolypeptideBuilder to build Polypeptide objects from Model and Chain objects as well.

OBTAINING THE SEQUENCE OF A STRUCTURE

The first thing to do is to extract all polypeptides from the structure (as above). The sequence of each polypeptide can then easily be obtained from the Polypeptide objects. The sequence is represented as a Biopython Seq object, and its alphabet is defined by a ProteinAlphabet object.

Example:

```
>>> seq = polypeptide.get_sequence()
>>> print(seq)
Seq('SNVVE...', <class Bio.Alphabet.ProteinAlphabet>)
```

ANALYZING STRUCTURES

Measuring Distances

The minus operator for atoms has been overloaded to return the distance between two atoms.

```
# Get some atoms
>>> ca1 = residue1['CA']
>>> ca2 = residue2['CA']
# Simply subtract the atoms to get their distance
>>> distance = ca1-ca2
```

Measuring Angles

Use the vector representation of the atomic coordinates, and the calc_ angle function from the Vector module:

```
>>> vector1 = atom1.get_vector()
>>> vector2 = atom2.get_vector()
>>> vector3 = atom3.get_vector()
>>> angle = calc_angle(vector1, vector2, vector3)
```

Measuring Torsion Angles

Use the vector representation of the atomic coordinates, and the calc_dihedral function from the Vector module:

```
>>> vector1 = atom1.get_vector()
>>> vector2 = atom2.get_vector()
>>> vector3 = atom3.get_vector()
>>> vector4 = atom4.get_vector()
>>> angle = calc_dihedral(vector1, vector2, vector3, vector4)
```

Determining Atom-Atom Contacts

Use NeighborSearch to perform neighbor lookup. The neighbor lookup is done using a KD tree module written in C (see Bio.KDTree), making it very fast. It also includes a fast method to find all point pairs within a certain distance of each other.

SUPERIMPOSING TWO STRUCTURES

Use a Superimposer object to superimpose two coordinate sets. This object calculates the rotation and translation matrix that rotates two lists of atoms on top of each other in such a way that their RMSD is minimized. Of course, the two lists need to contain the same number of atoms. The Superimposer object can also apply the rotation/translation to a list of atoms. The rotation and translation are stored as a tuple in the rotran attribute of the Superimposer object (note that the rotation is right multiplying!). The RMSD is stored in the rmsd attribute.

The algorithm used by Superimposer comes from Golub & Van Loan and makes use of singular value decomposition (this is implemented in the general Bio.SVDSuperimposer module).

Example:

```
>>> sup = Superimposer()
# Specify the atom lists
# 'fixed' and 'moving' are lists of Atom objects
```

```
# The moving atoms will be put on the fixed atoms
>>> sup.set_atoms(fixed, moving)
# Print rotation/translation/rmsd
>>> print(sup.rotran)
>>> print(sup.rms)
# Apply rotation/translation to the moving atoms
>>> sup.apply(moving)
```

To superimpose two structures based on their active sites, use the active site atoms to calculate the rotation/translation matrices (as above), and apply these to the whole molecule.

Mapping the Residues of Two Related Structures onto Each Other

First, create an alignment file in FASTA format, then use the StructureAlignment class. This class can also be used for alignments with more than two structures.

Calculating the Half Sphere Exposure

Half Sphere Exposure (HSE) is a new, 2D measure of solvent exposure [20]. Basically, it counts the number of Cα atoms around a residue in the direction of its side chain, and in the opposite direction (within a radius of 13 Å). Despite its simplicity, it outperforms many other measures of solvent exposure.

HSE comes in two flavors: HSEα and HSEβ. The former only uses the Cα atom positions, while the latter uses the Cα and Cβ atom positions. The HSE measure is calculated by the HSExposure class, which can also calculate the contact number. The latter class has methods which return dictionaries that map a Residue object to its corresponding HSEα, HSEβ and contact number values.

Example:

```
>>> model = structure[0]
>>> hse = HSExposure()
# Calculate HSEalpha
```

```
>>> exp_ca = hse.calc_hs_exposure(model, option='CA3')
# Calculate HSEbeta
>>> exp_cb=hse.calc_hs_exposure(model, option='CB')
# Calculate classical coordination number
>>> exp_fs = hse.calc_fs_exposure(model)
# Print HSEalpha for a residue
>>> print(exp_ca[some_residue])
```

DETERMINING THE SECONDARY STRUCTURE

For this functionality, you need to install DSSP (and obtain a license for it — free for academic use, see http://www.cmbi.kun.nl/gv/dssp/). Then use the DSSP class, which maps Residue objects to their secondary structure (and accessible surface area). The DSSP codes are listed in Table 7.1. Note that DSSP (the program, and thus by consequence the class) cannot handle multiple models!

Table 8.1: Codes employed for determining secondary structure in DSSP are described.

Code	Secondary Structure
H	α-helix
B	Isolated β-bridge residue
E	Strand
G	3-10 helix
I	Π-helix
T	Turn
S	Bend
-	Other

The DSSP class can also be used to calculate the accessible surface area of a residue.

CALCULATING THE RESIDUE DEPTH

Residue depth is the average distance of a residue's atoms from the solvent accessible surface. It's a fairly new and very powerful parameterization of solvent accessibility. For this functionality, you need to install Michel Sanner's MSMS program (http://www.scripps.edu/pub/olson-web/people/sanner/html/msms_home.html). Then use the ResidueDepth class. This class behaves as a dictionary which maps Residue objects to corresponding (residue depth, C^{α} depth) tuples. The C^{α} depth is the distance of a residue's C^{α} atom to the solvent accessible surface.

Example:

```
>>> model = structure[0]
>>> rd = ResidueDepth(model, pdb_file)
>>> residue_depth, ca_depth=rd[some_residue]
```

You can also get access to the molecular surface itself (via the get_surface function), in the form of a Numeric Python array with the surface points.

COMMON PROBLEMS IN PDB FILES

It is well known that many PDB files contain semantic errors (not the structures themselves, but their representation in PDB files). Bio.PDB tries to handle this in two ways. The PDBParser object can behave in two ways: a restrictive way and a permissive way, which is the default.

Example:

```
# Permissive parser
>>> parser = PDBParser(PERMISSIVE=1)
>>> parser = PDBParser() # The same (default)
# Strict parser
>>> strict_parser = PDBParser(PERMISSIVE=0)
```

In the permissive state (DEFAULT), PDB files that obviously contain errors are "corrected" (i.e. some residues or atoms are left out). These errors include:

- Multiple residues with the same identifier
- Multiple atoms with the same identifier (taking into account the altloc identifier)

These errors indicate real problems in the PDB file (for details see Hamelryck and Manderick, 2003]). In the restrictive state, PDB files with errors cause an exception to occur. This is useful to find errors in PDB files.

Some errors however are automatically corrected. Normally each disordered atom should have a non-blank altloc identifier. However, there are many structures that do not follow this convention, and have a blank and a non-blank identifier for two disordered positions of the same atom. This is automatically interpreted in the right way.

Sometimes a structure contains a list of residues belonging to chain A, followed by residues belonging to chain B, and again followed by residues belonging to chain A, i.e. the chains are 'broken'. This is also correctly interpreted.

Examples

Three exceptions were generated in cases where an unambiguous data structure could not be built. In all three cases, the likely cause is an error in the PDB file that should be corrected. Generating an exception in these cases is much better than running the chance of incorrectly describing the structure in a data structure.

Duplicate residues

One structure contains two amino acid residues in one chain with the same sequence identifier (resseq 3) and icode. Upon inspection it was found that this chain contains the residues Thr A3, …, Gly A202, Leu A3, Glu A204. Clearly, Leu A3 should be Leu A203. A couple of similar situations exist for structure 1FFK (which e.g. contains Gly B64, Met B65, Glu B65, Thr B67, i.e. residue Glu B65 should be Glu B66).

Duplicate atoms

Structure 1EJG contains a Ser/Pro point mutation in chain A at position 22. In turn, Ser 22 contains some disordered atoms. As expected, all atoms belonging to Ser 22 have a non-blank altloc specifier (B or C).

All atoms of Pro 22 have altloc A, except the N atom which has a blank altloc. This generates an exception, because all atoms belonging to two residues at a point mutation should have non-blank altloc. It turns out that this atom is probably shared by Ser and Pro 22, as Ser 22 misses the N atom. Again, this points to a problem in the file: the N atom should be present in both the Ser and the Pro residue, in both cases associated with a suitable altloc identifier.

Automatic correction

Some errors are quite common and can be easily corrected without much risk of making a wrong interpretation. These cases are listed below.

A blank **altloc** *for a disordered atom*

Normally each disordered atom should have a non-blank altloc identifier. However, there are many structures that do not follow this convention, and have a blank and a non-blank identifier for two disordered positions of the same atom. This is automatically interpreted in the right way.

Broken chains

Sometimes a structure contains a list of residues belonging to chain A, followed by residues belonging to chain B, and again followed by residues belonging to chain A, i.e. the chains are "broken." This is correctly interpreted.

Fatal errors

Sometimes a PDB file cannot be unambiguously interpreted. Rather than guessing and risking a mistake, an exception is generated, and the user is expected to correct the PDB file. These cases are listed below.

Duplicate residues

All residues in a chain should have a unique id. This id is generated based on:

- The sequence identifier (resseq).
- The insertion code (icode).

- The hetfield string ("W" for waters and "H_" followed by the residue name for other hetero residues)
- The residue names of the residues in the case of point mutations (to store the Residue objects in a DisorderedResidue object).

If this does not lead to a unique id something is quite likely wrong, and an exception is generated.

Duplicate atoms

All atoms in a residue should have a unique id. This id is generated based on:

- The atom name (without spaces, or with spaces if a problem arises).
- The *altloc* specifier.

If this does not lead to a unique id something is quite likely wrong, and an exception is generated.

ACCESSING THE PROTEIN DATA BANK

Downloading Structures from the Protein Data Bank

Structures can be downloaded from the PDB (Protein Data Bank) by using the retrieve_pdb_file method on a PDBList object. The argument for this method is the PDB identifier of the structure.

```
>>> pdbl = PDBList()
>>> pdbl.retrieve_pdb_file('1FAT')
```

The PDBList class can also be used as a command-line tool:

```
python PDBList.py 1fat
```

The downloaded file will be called pdb1fat.ent and stored in the current working directory. Note that the retrieve_pdb_file method also has an optional argument pdir that specifies a specific directory in which to store the downloaded PDB files.

The retrieve_pdb_file method also has some options to specify the compression format used for the download, and the program used for local decompression (default .Z format and gunzip). In addition, the PDB ftp site can be specified upon creation of the PDBList object. By default, the server of the Worldwide Protein Data Bank (ftp://ftp.wwpdb.org/pub/pdb/data/structures/divided/pdb/) is used. See the API documentation for more details. Thanks again to Kristian Rother for donating this module.

Downloading the Entire Pdb

The following commands will store all PDB files in the /data/pdb directory:

```
python PDBList.py all /data/pdb
python PDBList.py all /data/pdb -d
```

The API method for this is called download_entire_pdb. Adding the -d option will store all files in the same directory. Otherwise, they are sorted into PDB-style subdirectories according to their PDB ID's. Depending on the traffic, a complete download will take 2-4 days.

Keeping A Local Copy Of The Pdb Up To Date

This can also be done using the PDBList object. One simply creates a PDBList object (specifying the directory where the local copy of the PDB is present) and calls the update_pdb method:

```
>>> pl = PDBList(pdb='/data/pdb')
>>> pl.update_pdb()
```

One can of course make a weekly cronjob out of this to keep the local copy automatically up-to-date.

PDBList has some additional methods that can be of use. The get_all_obsolete method can be used to get a list of all obsolete PDB entries. The changed_this_week method can be used to obtain the entries that were added, modified or obsoleted during the current week. For more info on the possibilities of PDBList.

REFERENCES

1. International Tables for Crystallography - *Volume F: Crystallography of biological macromolecules*. Edited by E. Arnold, D. M. Himmel and M. G. Rossmann (**2012**).
2. Pauling L & Corey RB. ." *Proceedings of the National Academy of Sciences of the United States of America*. **37** (11): 251–256 (**1951**).
3. Pauling L, Corey RB & Branson HR. ." *Proceedings of the National Academy of Sciences of the United States of America*. **37** (11): 205–211 (**1951**).
4. Ramachandran GN., Ramakrishnan C. & Sasisekharan V. "Stereochemistry of Polypeptide Chain Configurations." *J. Mol. Biol.,* **7**, 95-99 (**1963**).
5. Blake CCF, Koenig DF, Mair GA, North ACT, Phillips DC & Sarma VR. "Structure of hen egg-white lysozyme. A three-dimensional Fourier synthesis at 2Å resolution." *Nature,* **206**, 757-761 (**1965**).
6. Kendrew JC, Dickerson RE, Strandberg BE, Hart RG, Davies DR, Phillips DC & Shore VC. "Structure Of Myoglobin: A Three-Dimensional Fourier Synthesis At 2Å Resolution." *Nature,* **185**, 422-427 (**1960**).
7. Perutz MF, Rossmann MG, Cullis AF, Muirhead H, Will G & North ACT. (**1960**) "Structure Of Haemoglobin: A Three-Dimensional Fourier Synthesis At 5.5Å. Resolution, Obtained By X-Ray Analysis." *Nature*, **185**, 416-422.
8. Cock PA, Antao T, Chang JT, Bradman BA, Cox CJ, Dalke A, Friedberg I, Hamelryck T, Kauff F, Wilczynski B and de Hoon MJL. "Biopython: freely available Python tools for computational molecular biology and bioinformatics." *Bioinformatics*, **25**, 1422-1423 (**2009**).

9

SEQUENCE - STRUCTURE COMPARISON: EVOLUTIONARY ANALYSIS

In the past chapters, you have learned many computational concepts of comprehending biological sequences and structures. In particular, you have leanred how to handle protein/DNA sequences, manipulate them, become familiarized with different sequence formats, and been introduced to how three-dimensional structural information is stored and distributed through publicly available databases. We also developed some simple recipes for common tasks in molecular analysis.

In this final chapter, we try to integrate the sequence of structure paradigms and apply it to the study of the evolution of biomolecular sequences. This is a standard problem in biology and we are going to develop and use a handy recipe for the thorough comprehension of this important concept.

Let us start with our computational requirements for this chapter.

- We will need two Python packages: **ProDy** and **Matplotlib**.
- We will be peforming our analysis in IPython mode. This comes along with Anaconda distribution. I highly recommend setting up your respective Anaconda distributions and environments.

Installation of these packages is straighforward and is along the lines of the packages you have previously installed in this book or elsewhere. Hence, we will get started in the next section.

ACCESSING PFAM

Pfam is a database of Protein Families. Pfam is a comprehensive collection of protein domains and families, represented as multiple sequence alignments and as profile hidden Markov models. Pfam was founded in 1995 by Erik Sonhammer, Sean Eddy and Richard Durbin as a collection of commonly occurring protein domains that could be used to annotate the protein coding genes of multicellular animals. Pfam families are divided into two categories, Pfam-A and Pfam-B. Each Pfam-A family consists of a curated seed alignment containing a small set of representative members of the family, profile hidden Markov models (profile HMMs) built from the seed alignment and an automatically generated full alignment which contains all detectable protein sequences belonging to the family, as defined by profile HMM searches of primary sequence databases. The use of HMMS accords greater weights to matches at conserved sites, allowing better remote homology detection. Pfam-B entries are automatically generated from the ProDom database, and are represented by a single alignment. The use of representative seed alignments for Pfam-A families allows efficient and sustainable manual curation of alignments and annotation, while the automatic generation of full alignments and Pfam-B clusters ensures that Pfam is a comprehensive classification of protein families that scales effectively with the growth of the sequence databases.

The availability of 3D protein structures has been essential for finding distant evolutionary relationships and understanding protein function at the molecular level. Sequences with a known structure usually have a clear domain organization and ideally each domain should be represented in Pfam. Pfam data are freely accessible via the web and are available for download in a variety of forms. The latest version released in March 2017, Pfam 31.0, contains 16,712 families.

We will show how to access the data from Pfam using ProDy package.

```
In [1]: from prody import *
from pylab import *
ion() # turn interactive mode ON
```

In this section we develop a recipe to access Pfam database We will perform our searching Pfam with UniProt ID of **PIWI_ARCFU** or PDB ID of **3lucA** (A stands for chain identifier) Matching Pfam accession (one or more) as keys will map to a dictionary that contains locations (alignment start, end, evalue etc), pfam family type, accession and id. We query Pfam using the *searchPfam()* with a UniProt ID.

```
In [2]: matches = searchPfam('PIWI_ARCFU')
@> Retrieving Pfam search results: http://pfam.xfam.org/protein/PIWI_ARCFU?output=xml
@> Pfam search completed in 1.03s.
@> Query 'PIWI_ARCFU' matched 1 Pfam families.
```

It is a good practice to save this record on disk, as NCBI may not respond to repeated searches for the same sequence. We can do this using Python standard library pickle as follows:

```
In [3]: import pickle
pickle.dump(matches, open('pfam_search_PIWI_ARCFU.pkl', 'w'))
matches
Out[3]:
{'PF02171': {'accession': 'PF02171',
 'id': 'Piwi',
 'locations': [{'ali_end': '405',
   'ali_start': '111',
   'bitscore': '238.30',
   'end': '406',
   'evalue': '1.6e-67',
   'hmm_end': '300',
   'hmm_start': '2',
   'start': '110'}],
 'type': 'Pfam-A'}}
```

Input can also be a protein sequence or a file containing the sequence. I leave it to the reader to figure out a way to do it using ProDy package. For sequence searches, we can pass additional parameters to *searchPfam()* like *search_b* which will search Pfam-B and *skip_a* that will not search Pfam-A database. Additional parameters include ga that uses gathering threshold instead of e-value, evalue cutoff can also be specified and timeout that can be set higher especially when searching larger sequences, default is *timeout* = 60 seconds.

One can also retrieve precalculated Multiple Sequence Alignment files from Pfam. The readers would develop their own MSA for specific research problems, we proceed in this section by retrieving MSA from Pfam. Data from Pfam database can be fetched using *fetchPfamMSA()*. Valid inputs are Pfam ID, e.g. *Piwi*, or Pfam accession, e.g. *PF02171* obtained from *searchPfam()*. Alignment type can be "*full'* (default), "*seed*", "*ncbi*" or "*metagenomics*" or "*rp15*" or "*rp35*" or "*rp55*" or "*rp75*." Here, *rp* stands for representative proteomes.

In [4]: fetchPfamMSA('piwi', alignment='seed')

@> Pfam MSA for piwi is written as .\piwi_seed.sth.

Out[4]:

'.\\piwi_seed.sth'

A compressed file can be downloaded by setting *compressed=True*. The format of the MSA can be of "*selex*" (default), "*stockholm*" or "*fasta*." This will return the path of the downloaded MSA file. The output name can be specified, for by default it will have "accession/ID_alignment.format".

Note that in this case we passed a folder name, the downloaded file is saved in this folder, after it is created if it did not exist. Also longer timeouts are necessary for larger families. Some other parameters like *gap*, *order* or *inserts* can be set, as shown in the following example.

In [5]: fetchPfamMSA('PF02171', compressed=True, gaps='mixed', inserts='lower', order='alphabetical', format='fasta')

@> Pfam MSA for PF02171 is written as .\PF02171_full.fasta.gz.

Out[5]:

'.\\PF02171_full.fasta.gz'

Let us look at the MSA and see how we can computationally *parse*, *refine*, *filter*, *slice* and *write* a modified MSA. Let us begin by getting Pfam MSA for protein family that contains **PIWI_ARCFU**.

```
In [6]: searchPfam('PIWI_ARCFU').keys()
fetchPfamMSA('PF02171', alignment='seed')
@> Retrieving Pfam search results: http://pfam.xfam.org/protein/PIWI_ARCFU?output=xml
@> Pfam search completed in 0.31s.
@> Query 'PIWI_ARCFU' matched 1 Pfam families.
@> Pfam MSA for PF02171 is written as .\PF02171_seed.sth.
Out[6]:
'.\\PF02171_seed.sth'
```

PARSING MSA FILES

Now that we have access to Pfam MSAs, we will be using the function *MSAFile()* or *parseMSA()* to read the MSA file. Reading using MSAFile yields an *MSAFile* object. Iterating over the object will yield an object of *Sequence* from which *labels*, *sequence* can be obtained.

```
In [7]: msafile = 'PF02171_seed.sth'
In [8]: msafobj = MSAFile(msafile)
In [9]: msafobj
Out[9]: <MSAFile: PF02171_seed (Stockholm; mode 'rt')>
In [10]: msa_seq_list = list(msafobj)
In [11]: msa_seq_list[0]
Out[11]: <Sequence: AGO1_SCHPO (length 395; 300 residues and 95 gaps)>
```

parseMSA() returns an MSA object. We can parse compressed files, but reading uncompressed files are much faster.

```
In [12]: fetchPfamMSA('PF02171', compressed=True)
@> Pfam MSA for PF02171 is written as .\PF02171_full.sth.gz.
Out[12]: '.\\PF02171_full.sth.gz'
In [13]: parseMSA('PF02171_full.sth.gz')
@> 5084 sequence(s) with 1752 residues were parsed in 0.25s.
```

```
Out[13]: <MSA: PF02171_full (5084 sequences, 1752 residues)>
In [14]: fetchPfamMSA('PF02171', format='fasta')
@> Pfam MSA for PF02171 is written as .\PF02171_full.fasta.
Out[14]: '.\\PF02171_full.fasta'
In [15]: parseMSA('PF02171_full.fasta.gz')
@> 5084 sequence(s) with 1752 residues were parsed in 0.61s.
Out[15]: <MSA: PF02171_full (5084 sequences, 1752 residues)>
```

Iterating over a file will yield sequence id, sequence, residue start and end indices:

```
In [16]: msa = MSAFile('PF02171_seed.sth')
for seq in msa:
    seq
```

Now, let us get into filtering and slicing of MSA. We take up Filtering first in the recipe. Any function that takes label and sequence arguments and returns a boolean value can be used for filtering the sequences. A sequence will be yielded if the function returns True. In the following example, sequences from organism **ARATH** are filtered.

```
In [17]: msafobj_ara = MSAFile(msafile, filter=lambda lbl, seq: 'ARATH' in lbl)
In [18]:
for seq in msafobj_ara:
    seq.getLabel()
```

Let us take slicing. A list of integers can be used to slice sequences as follows. This enables selective parsing of the MSA file.

```
In [19]: msafobj_slic = MSAFile(msafile, slice=list(range(10)) + list(range(374,384)))
In [20]: list(msafobj_slic)[0]
Out[20]: <Sequence: AGO1_SCHPO (length 20; 19 residues and 1 gaps)>
```

The MSA object offers other functionalities as well. These include querying, indexing, slicing row and columns and refinement.

writing msa

We now look at how to write MSA. The function writeMSA() can be used to write MSA. It takes filename as input which should contain appropriate extension that can be ".slx" or ".sth" or ".fasta" or format should be specified as "SELEX", "Stockholm" or "FASTA." Input MSA should be MSAFile or MSA object. Filename can contain ".gz" extension, in which case a compressed file will be written.

In [21]: writeMSA('sliced_MSA.fasta', msafobj)

Out[21]: 'sliced_MSA.fasta'

writeMSA() returns the name of the MSA file that is written.

EVOLUTIONARY ANALYSIS

Let us now step into Evolutionary Analysis using ProDy. We take up protein family **RnaseA** for this analysis. Although, the same can be done using any other protein family of interest, we use this example for its ease and simplicity. Let's download full MSA file for the **Rnase family**. We can do this by specifying the PDB ID of a protein in the family (for example **1K2A**).

In [22]: searchPfam('1K2A').keys()

@> PDB file is found in working directory (1k2a.pdb.gz).

@> UniProt ID code RNAS2_HUMAN for 1K2A chain A will be used.

@> Retrieving Pfam search results: http://pfam.xfam.org/protein/RNAS2_HUMAN?output=xml

@> Pfam search completed in 0.50s.

@> Query '1K2A' matched 1 Pfam families.

Out[22]: ['PF00074']

In [23]: fetchPfamMSA('PF00074')

@> Pfam MSA for PF00074 is written as .\PF00074_full.sth.

Out[23]: '.\\PF00074_full.sth'

In [24]: msa_rnase = parseMSA('PF00074_full.sth')

@> 645 sequence(s) with 248 residues were parsed in 0.00s.

The MSA has 645 sequences and 248 residues. We are going to refine the MSA to decrease the number of gaps in the alignment. We will remove any columns in the alignment for which there is a gap in the specified PDB file, and then remove any rows that have more than 20% gaps. *refineMSA()* does all of this and returns an MSA object.

In [25]: msa_rnase_refine = refineMSA(msa_rnase, label='RNAS2_HUMAN', rowocc=0.8, seqid=0.98)

@> Label refinement reduced number of columns from 248 to 126 in 0.00s.

@> Row occupancy refinement reduced number of rows from 645 to 493 in 0.00s.

@> Sequence identity refinement reduced number of rows from 493 to 420 in 0.08s.

In [26]: msa_rnase_refine

Out[26]: <MSA: PF00074_full refined (label=RNAS2_HUMAN, rowocc>=0.8, seqid>=0.98) (420 sequences, 126 residues)>

MSA is refined based on the sequence of RNAS2_HUMAN, corresponding to 1K2A. The refined MSA has only 420 sequences and 126 residues.

Let us try to understand the calculation of **occupancy** for each column in MSA. This helps us to figure out those columns which harbour relatively larger gaps in a MSA. ProDy provides a function, *showMSAOccupancy()*, for visually represnting the calculations of MSA occupancy. It uses the *calcMSAOccupancy()* function, which evaluates the occupancy for a MSA. All the plotting functions have a prefix '*show*'.

In [27]: %matplotlib inline // a command to visualize plots in IPython notebook environment

rcParams['figure.figsize'] = (15, 8) // set parameters of figure size

showMSAOccupancy(msa_rnase_refine, occ='res')

Out[27]: <Container object of 126 artists>

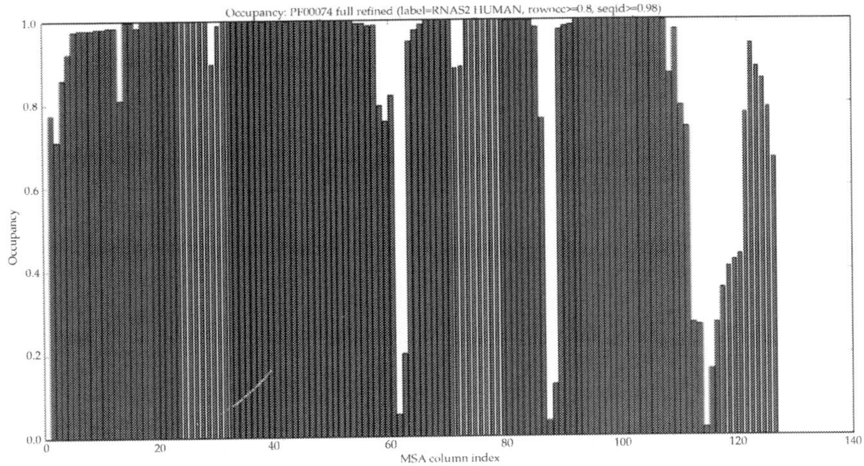

Let's find the minimum. This can be achieved by using the min() function of Python:

In [28]: calcMSAOccupancy(msa_rnase_refine, occ='res').min();

We can also specify indices based on the PDB.

In [29]: ind_pdb = list(range(4,130))

In [30]: showMSAOccupancy(msa_rnase_refine, occ='res', indices=ind_pdb);

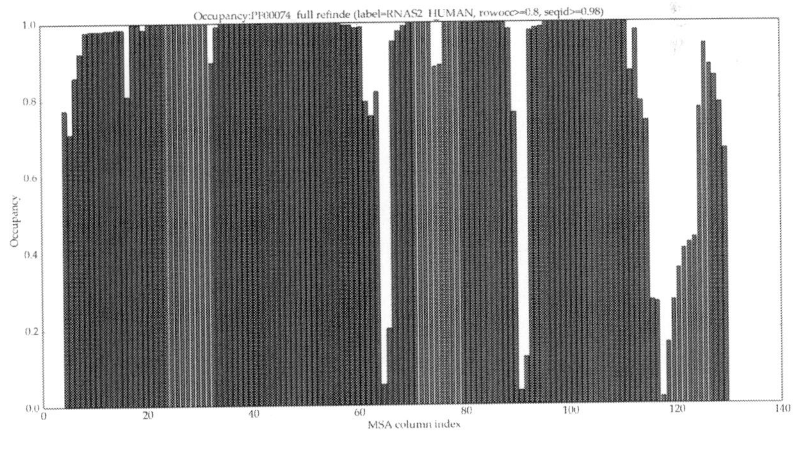

ENTROPY CALCULATION

Calculating and plotting **Shannon Entropy** for each position in MSA. ProDy provides a function for achieving this using *calcShannonEntropy()*. It takes the MSA object as input and returns a 1D numpy array.

```
In [31]: entropy_rnase = calcShannonEntropy(msa_rnase_refine)
In [32]: entropy_rnase
Out[32]:
array([2.36613736, 1.99470176, 1.65364808, 1.88917136, 2.02235343,
       0.39425203, 2.06096278, 1.91402193, 1.08894235, 0.53820681,
       1.60139711, 1.60724268, 2.4537103 , 2.31404239, 2.43073651,
       2.57425229, 1.44189669, 0.06370385, 1.2705884 , 2.68715206,
       1.67040717, 0.57643524, 2.29807206, 2.3664016 , 1.42974535,
       1.84936187, 2.17878595, 1.91218315, 2.49939829, 2.54707541,
       2.47515606, 0.12276959, 0.73511379, 2.39556536, 2.31974871,
       0.61887565, 0.92512854, 0.36234225, 1.05908081, 0.94919649,
       1.93072709, 2.02880476, 2.27988375, 2.54893929, 2.17843096,
       1.10039053, 2.25588644, 2.0483008 , 1.543456  , 0.07714001,
       2.55287734, 2.39489292, 2.2433411 , 2.08416789, 2.07547272,
       2.44361436, 1.15860861, 1.54296911, 1.06679809, 1.73017681,
       1.61795071, 1.90365599, 2.6019284 , 2.56214442, 1.48540281,
       0.9738367 , 1.56800427, 2.27912331, 0.67976519, 2.55232234,
       2.35623829, 2.11072283, 1.39357801, 2.34532866, 1.58169238,
       0.73468154, 2.3100801 , 0.03021297, 2.11704277, 1.43502813,
       1.80529625, 2.02027233, 1.73296964, 2.12547756, 2.37043995,
       2.40121641, 0.72987121, 1.61041162, 1.17953975, 2.07238397,
       0.03051912, 2.14556677, 0.08044071, 2.075509  , 2.21154746,
       2.25248083, 2.58758523, 2.50007037, 2.54719547, 1.82228061,
       2.25981663, 1.53649634, 1.856788  , 1.03331281, 1.28119847,
      -0.        , 1.97376825, 2.15221636, 2.31902459, 2.09233594,
```

2.47500418, 2.04629137, 0.43502974, -0. , 1.2017231 ,
1.23664303, 2.28258955, 1.5759423 , 1.9509346 , 1.53194984,
1.40660243, 0.06214533, 0.50995907, 0.49144669, 1.05592875,
0.14211381])

Now, let us plot this using *showShannonEntropy()* function.

In [33]: showShannonEntropy(entropy_rnase, ind_pdb)

Out[33]: <Container object of 126 artists>

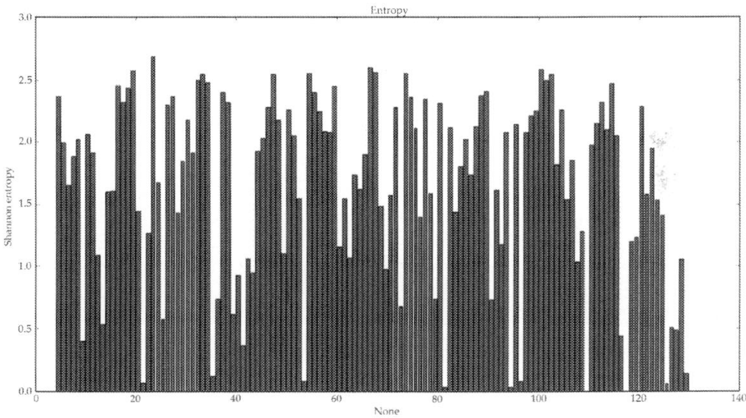

MUTUAL INFORMATION

We can also calculate mutual information between the positions of MSA using *buildMutinforMatrix()* which also takes MSA object as input.

In [34]: mutinfo_rnase = buildMutinfoMatrix(msa_rnase_refine)

@> Mutual information matrix was calculated in 0.06s.

In [35]: mutinfo_rnase

Out[35]:

array([[0. , 0.6341196 , 0.67416245, ..., 0.22796202,
0.36778644, 0.13761952],
[0.6341196 , 0. , 0.68775638, ..., 0.29381926,
0.36364212, 0.20952131],
[0.67416245, 0.68775638, 0. , ..., 0.36268407,

```
  0.43242045, 0.25035628],
 ...,
 [ 0.22796202, 0.29381926, 0.36268407, ..., 0.        ,
   0.4691457 , 0.27946279],
 [ 0.36778644, 0.36364212, 0.43242045, ..., 0.4691457 ,
   0.        , 0.32909111],
 [ 0.13761952, 0.20952131, 0.25035628, ..., 0.27946279,
   0.32909111, 0.        ]])
```

This is a 2D Numpy array object. We can apply normalization and correction to this matrix using the functions *applyMutinforNorm()* and *applyMutinfoCorr()*. Let us see how this is done for our example.

In [36]: mutinfo_norm = applyMutinfoNorm(mutinfo_rnase, entropy_rnase, norm='minent')

In [37]: mutinfo_corr = applyMutinfoCorr(mutinfo_rnase, corr='apc')

Now, we get to see the mutual information matrix with the above effects. We also play around with a few parameters to see how these effects are modified.

In [38]: showMutinfoMatrix(mutinfo_rnase)

Out[38]: (<matplotlib.image.AxesImage at 0xa151080>,
<matplotlib.colorbar.Colorbar at 0xa2125f8>)

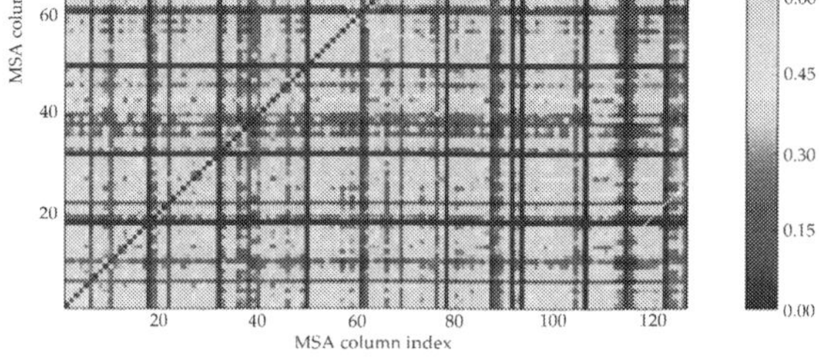

In [39]: showMutinfoMatrix(mutinfo_corr, clim=[0, mutinfo_corr.max()], xlabel='1K2A: 4-131')

Out[39]: (<matplotlib.image.AxesImage at 0xc00c550>, <matplotlib.colorbar.Colorbar at 0xc0d2518>)

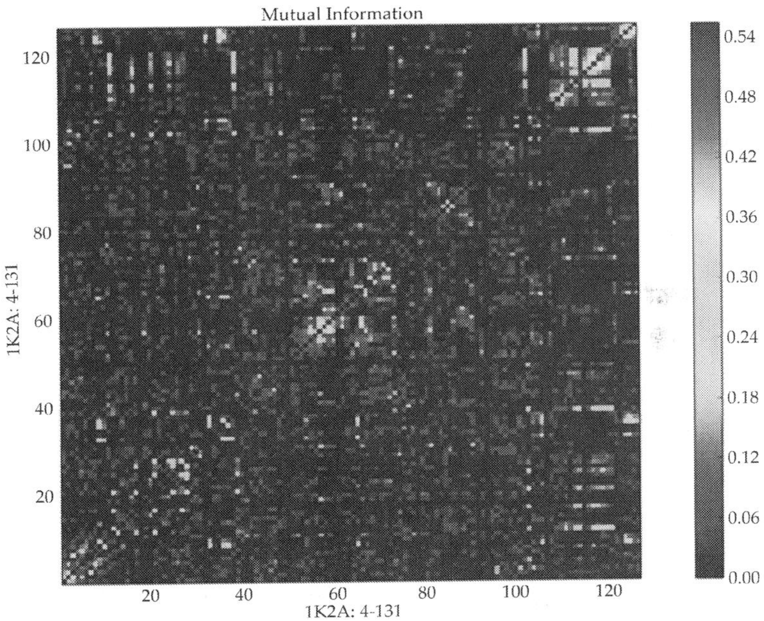

You may want to save the mutual information and entropy arrays to a file for your reference. In which case, you can use the *writeArray()* function to write Numpy array data to a file.

In [40]: writeArray('1K2A_MI.txt', mutinfo_rnase)

Out[40]: '1K2A_MI.txt'

RANK ORDERING

You can analyze further by rank ordering the matrix and analyzing the pairs with highest mutual information or the most co-evolving residues. You can achieve this using the **ProDy** function *calcRankOrder()*. A z-score normalization can also be applied to select co-evolving pairs based on a z-score cut-off.

```
In [41]: rank_row, rank_col, zscore_sort = calcRankorder(mutinfo_rnase, zscore=True)
@> Zscore normalization has been applied.
@> Matrix is symmetric, only lower triangle indices will be returned.
In [42]: #indices = list(range(4,132))
In [43]: asarray(ind_pdb)[rank_row[:5]]
Out[43]: array([126, 31, 63, 120, 30])
In [44]: asarray(ind_pdb)[rank_col[:5]]
Out[44]: array([125, 21, 62, 118, 25])
In [45]: zscore_sort[:5]
Out[45]: array([ 3.15363221, 3.11856249, 3.02581626, 2.88948586, 2.81237362])
```

Sequence - Stucture Comparison

We will delve into sequence conservation properties with structural mobility obtained from **Gaussian Network Model (GNM)** calculations. We will spent some time in understanding the concept of GNM, before taking up a recipe.

GNM is a representation of a biomolecule as an elastic mass-and-spring network to study and characterize the mechanical aspects of its long-term large-scale dynamics. In a GNM the protein is represented by nodes, corresponding to C^{α} of amino acid residues. Most of the biological phenomena, for example enzyme catalysis, occur in a range of timescales from nano- to milliseconds. Molecular dynamics simulations rarely reach microsecond trajectory length. But, GNM can access biomolecular motions at such timescales. The model captures native state functional motions of a biomolecule. A schematic illustrating the concept is given below.

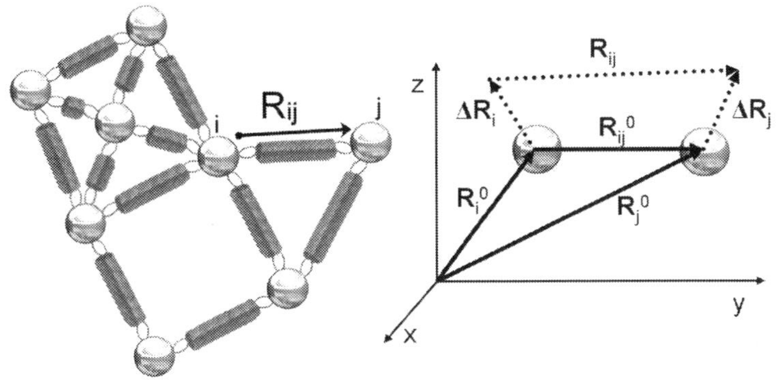

Figure 9.1: Schematic representation of nodes in elastic network of GNM. Every node is connected to its spatial neighbors by uniform springs. Distance vector between two nodes, **i** and **j**, is shown by an arrow and labeled $\mathbf{R}_{ij}$. Equilibrium positions of the **i**th and **j**th nodes, $\mathbf{R}^0_i$ and $\mathbf{R}^0_j$, are shown in xyz coordinate system. $\mathbf{R}^0_{ij}$ is the equilibrium distance between nodes **i** and **j**. Instantaneous fluctuation vectors, $\Delta\mathbf{R}_i$ and $\Delta\mathbf{R}_j$, and instantaneous distance vector, $\mathbf{R}_{ij}$, are shown by the dashed arrows.

The GNM was proposed by Bahar, Atilgan, Haliloglu and Erman in 1997. It was motivated by a study by I. Tirion, where he showed by normal mode analysis that the representation of atomic interactions by harmonic potentials with uniform spring constant (instead of detailed force field functions and parameters) would not alter the most collective modes of motions predicted at the lowest frequency end of the spectrum, in accord with earlier works that utilized normal mode analysis and simplified harmonic potentials to study the dynamics. The theory is quite involved and does not fit within the scope of the present book. Interested readers are directed to the papers by Bahar et al., and hat of Tirion.

The GNM differs from normal mode analyses and simulations, as it offers an analytical formulation and unique solution for each structure, exclusively based on inter-residue contact topology.

The dynamics of the interactions is controlled by the *connectivity* (or Kirchhoff) *matrix* **G**, by analogy with the statistical mechanical theory of elasticity originally developed by Flory and coworkers for polymer networks. The elements of **G** are defined as

This can be considered as a matrix of *inter-residue contacts*. Here r_c is the cutoff distance defining the range of interaction of residues, each residue being represented by its a-carbon, and $\mathbf{R}_{ij}$ is the distance between ith and jth residues. The value of r_c = **7 Å includes the neighboring residues located in the first coordination shell near a central residue.**

The determinant of $\mathbf{\Gamma}$ is 0, and hence $\mathbf{\Gamma}^{-1}$ cannot be calculated directly. Instead, it is found from the eigenvalue decomposition $\mathbf{\Gamma} = \mathbf{ULU}^T$ which results in an ensemble of N-1 independent modes. **U** is the orthogonal matrix whose k^{th} column $\boldsymbol{u}_k$ is the kth **eigenvector** of **Γ,** and **L** is the diagonal matrix of **eigenvalues** l_k which are usually organized in ascending order. One of the eigenvalues is identically zero, and the remaining N-1 eigenvalues define each the frequency of the N-1 GNM modes. The ith element $(\boldsymbol{u}_k)_i$ of $\boldsymbol{u}_k$ describes the motion of the ith residue along the kth normal mode coordinate; or, the elements of the kth eigenvector $\boldsymbol{u}_k$ represent the distribution of residue displacements (normalized over all residues) along the kth mode axis, and the corresponding eigenvalue l_k **scales with the frequency of the mode. The size of the motion along the** k^{th} mode scales with $(1/l_k)^{1/2}$. Thus $(1/l_k)^{1/2}$ serves as the 'weight' of a displacement along mode k. Clearly the lowest frequency modes (small l_k) also called *slowest* modes or *softest* modes make the largest contribution to the overall motion.

The cross-correlation between residues can be written as a sum of N-1 GNM modes as

The **mean-square fluctuations** of residue i in mode k can be evaluated from Eq. 6 by replacing j by i. The plot of $tr[\boldsymbol{u}_k \boldsymbol{u}_k^T]$ as a function of residue i represents the probability distribution of residue square fluctuations in mode k, also called k^{th} *mode profile.*

The cross-correlations are usually shown in a matrix/map format, with the diagonal terms representing the mean-square fluctuations. This ij^{th} element of this matrix, termed cross-correlation matrix, is

$\mathbf{C}_{ij}$ =

Let us see how we can achieve the above theoretical concepts computationally.

In [46]:

Assigning the previously parsed MSA msa_rnase to a new variable for the present recipe

msa_seq_str = msa_rnase

We refine this MSA based on the sequence of RNAS1_BOVIN

In [47]: msa_seq_str_refine = refineMSA(msa_seq_str, label='RNAS1_BOVIN', rowocc=0.8, seqid=0.98)

@> Label refinement reduced number of columns from 248 to 119 in 0.00s.

@> Row occupancy refinement reduced number of rows from 645 to 583 in 0.00s.

@> Sequence identity refinement reduced number of rows from 583 to 485 in 0.09s.

Next, we calculate the entropy for the refined MSA:

In [48]: entropy_seq_str = calcShannonEntropy(msa_seq_str_refine)

Mobility Calculation

We want to understand the residue fluctuations or mobility of residues for a protein member of the above family. We will use chain B of **2W5I**.

In [49]: pdb = parsePDB('2W5I', chain='B')

@> PDB file is found in working directory (2w5i.pdb.gz).

@> 973 atoms and 1 coordinate set(s) were parsed in 0.02s.

In [50]: chB_ca = pdb.select('protein and name CA and resid 1 to 119')

We perform GNM calculation by invoking *GNM()* function as follows:

In [51]: gnm = GNM('2W5I')

In [52]: gnm.buildKirchhoff(chB_ca)

@> Kirchhoff was built in 0.01s.

In [53]: gnm.calcModes(n_modes=None)

@> 118 modes were calculated in 0.01s.

Now, one can obtain residue mobility using slowest mode. We can also check mobilty for slowest 5 modes and / or all modes.

In [54]: mobility_1 = calcSqFlucts(gnm[0])

In [55]: mobility_1to8 = calcSqFlucts(gnm[:8])

In [56]: mobility_all = calcSqFlucts(gnm[:])

Compare mobility and conservation

Now, we are in a position to compare structural mobility with degree of conservation. We do this by calculating correlation coefficient between the two quantities.

```
In [57]: #len(mobility_all)
result = corrcoef(mobility_all, entropy_seq_str)
In [58]: result.round(3)[0,1]
Out[58]: 0.085000000000000006
In [59]: indices_comp = range(1,120)
In [60]: xlim(min(indices_comp)-1, max(indices_comp)+1);
bar(indices_comp, entropy_seq_str, width=1.2, color='grey', hold='True');
plot(indices_comp,mobility_all*(max(entropy_seq_str)/mean(mobility_all)), color='b', linewidth=2);
```

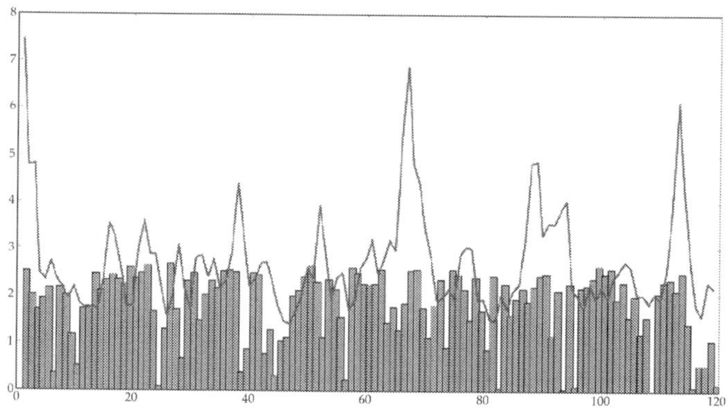

The entropy reflecting the conservation of each residues is shown as bar, whereas the mobility is represented as a blue line above the bar. This brings out the fact that most conserved residues, ordered residues , are least mobile structurally. One can see that peaks appear at residues with high Shannon entropy.

SUMMARY

In this chapter, you have learnt:

- How to computationally access Pfam database
- Retrive precompiled MSA for protein family of interest
- Write a refined MSA
- Perform evolutionary analysis of RnaseA family of proteins usinf **ProDy** package.
- Understood the concept of Gaussian Network Model for biomolecules, which helps in comprehnding the functionally significant motions of proteins.

REFERENCES

1. Sonnhammer, Erik L.L.; Eddy, Sean R.; Durbin, Richard. "Pfam: A Comprehensive Database of Protein Domain Families Based on Seed Alignments." *Proteins*. **28**: 405–420 (**1997**).
2. Finn RD, Tate J, Mistry J, Coggill PC, Sammut SJ, Hotz HR, Ceric G, Forslund K, Eddy SR, Sonnhammer EL, Bateman A. "The Pfam protein families database." *Nucleic Acids Res.* **36** (Database issue): D281–8 (**2008**).
3. Bru C, Courcelle E, Carrère S, Beausse Y, Dalmar S, Kahn D " The ProDom database of protein domain families: more emphasis on 3D." *Nucleic Acids Res.* Jan 1; **33**(Database issue):D212-5 (**2005**).
4. Bakan A, Meireles LM, Bahar I (2011). "ProDy: Protein Dynamics Inferred from Theory and Experiments." *Bioinformatics* **27**(11): 1575-1577 (**2011**).
5. Bahar I, Atilgan AR, Erman B. Direct evaluation of thermal fluctuations in protein using a single parameter harmonic potential. *Folding & Design* **2**:173-181 (**1997**).
6. Haliloglu T, Bahar I, Erman B. "Gaussian dynamics of folded proteins." *Phys. Rev.Lett.* **79**:3090-3093 (**1997**).
7. https://en.wikipedia.org/wiki/Gaussian_network_model
8. Tirion, M.M. "Large amplitude elastic motions in proteins from a single-parameter, atomic analysis." *Phys. Rev. Lett.* **77**: 1905 (**1996**).

APPENDICES

Appendix 1

We present here a collection of useful regular expressions. These will be useful in your working with Python. In addition, it would be fun and good for learning to harness the power of regular expressions.

The complete documentation of the Python re module can be found here:

https://docs.python.org/3.4/howto/regex.html

Below, I just want to list the most important methods for convenience:

re.match() : Determine if the RE matches at the beginning of the string.

re.search() : Scan through a string, looking for any location where this RE matches.

re.findall() : Find all substrings where the RE matches, and returns them as a list.

re.finditer() : Find all substrings where the RE matches, and returns them as an iterator.

If you are using the same regular expression multiple times, it is recommended to compile it for improved performance.

compiled_re = re.compile(r'some_regexpr')

for word in text:

match = comp.search(compiled_re))

do something with the match

Let us take a few examples for illustration:

```
import re
needle = 'needlers'
# Python approach
print(bool(any([needle.endswith(e) for e in ('ly', 'ed', 'ing', 'ers')])))
# On-the-fly Regular expression in Python
print(bool(re.search(r'(?:ly|ed|ing|ers)$', needle)))
# Compiled Regular expression in Python
comp = re.compile(r'(?:ly|ed|ing|ers)$')
print(bool(comp.search(needle)))
True
True
True
```

1. Identify files via file extensions

A regular expression to check for file extensions.

This approach is not recommended for thorough limitation of file types (parse the file header instead). However, this *regex* is still a useful alternative to e.g., a Python's *endswith* approach for quick pre-filtering for certain files of interest.

```
pattern = r'(?i)(\w+)\.(jpeg|jpg|png|gif|tif|svg)$'
# remove `(?i)` to make regexpr case-sensitive
str_true = ('test.gif', 'image.jpeg', 'image.jpg', 'image.TIF')
str_false = ('test.pdf', 'test.gif.pdf',)
for t in str_true:
   assert(bool(re.match(pattern, t)) == True), '%s is not True' %t
for f in str_false:
   assert(bool(re.match(pattern, f)) == False), '%s is not False' %f
```

2. Username validation

Checking for a valid user name that has a certain minimum and maximum length.

Allowed characters:

letters (upper- and lower-case)

numbers

dashes

underscores

```
min_len = 5 # minimum length for a valid username
max_len = 15 # maximum length for a valid username
pattern = r"^(?i)[a-z0-9_-]{%s,%s}$" %(min_len, max_len)
# remove `(?i)` to only allow lower-case letters
str_true = ('user123', '123_user', 'Username')
str_false = ('user', 'username1234_is-way-too-long', 'user$34354')
for t in str_true:
   assert(bool(re.match(pattern, t)) == True), '%s is not True' %t
```

```
for f in str_false:

    assert(bool(re.match(pattern, f)) == False), '%s is not False' %f
```

3. Checking for valid email addresses

A regular expression that captures most email addresses.

```
pattern = r"(^[a-zA-Z0-9_.+-]+@[a-zA-Z0-9-]+\.[a-zA-Z0-9-.]+$)"

str_true = ('test@mail.com',)

str_false = ('testmail.com', '@testmail.com', 'test@mailcom')

for t in str_true:

    assert(bool(re.match(pattern, t)) == True), '%s is not True' %t

for f in str_false:

    assert(bool(re.match(pattern, f)) == False), '%s is not False' %f
```

4. Check for a valid URL

Checks for an URL if a string ...

starts with *https://*, or *http://*, or *www.* or ends with a *dot* extension

```
pattern = '^(https?:\/\/)?([\da-z\.-]+)\.([a-z\.]{2,6})([\/\w \.-]*)*\/?$'

str_true = ('https://github.com', 'http://github.com', 'www.github.com', 'github.com', 'test.de', 'https://github.com/rasbt', 'test.jpeg' # !!!)

str_false = ('testmailcom', 'http:testmailcom', )

for t in str_true:

    assert(bool(re.match(pattern, t)) == True), '%s is not True' %t

for f in str_false:

    assert(bool(re.match(pattern, f)) == False), '%s is not False' %f
```

5. Checking for numbers

Positive integers

```
pattern = '^\d+$'

str_true = ('123', '1', )

str_false = ('abc', '1.1', )

for t in str_true:

    assert(bool(re.match(pattern, t)) == True), '%s is not True' %t

for f in str_false:

    assert(bool(re.match(pattern, f)) == False), '%s is not False' %f
```

Negative integers

```
pattern = '^-\d+$'

str_true = ('-123', '-1', )

str_false = ('123', '-abc', '-1.1', )

for t in str_true:

    assert(bool(re.match(pattern, t)) == True), '%s is not True' %t

for f in str_false:

    assert(bool(re.match(pattern, f)) == False), '%s is not False' %f
```

All integers

```
pattern = '^-{0,1}\d+$'
str_true = ('-123', '-1', '1', '123',)
str_false = ('123.0', '-abc', '-1.1', )
for t in str_true:
    assert(bool(re.match(pattern, t)) == True), '%s is not True' %t
for f in str_false:
    assert(bool(re.match(pattern, f)) == False), '%s is not False' %f
```

Positive numbers

```
pattern = '^\d*\.{0,1}\d+$'
str_true = ('1', '123', '1.234', )
str_false = ('-abc', '-123', '-123.0')
for t in str_true:
    assert(bool(re.match(pattern, t)) == True), '%s is not True' %t
for f in str_false:
    assert(bool(re.match(pattern, f)) == False), '%s is not False' %f
```

Negative numbers

```
pattern = '^-\d*\.{0,1}\d+$'
str_true = ('-1', '-123', '-123.0', )
str_false = ('-abc', '1', '123', '1.234', )
for t in str_true:
    assert(bool(re.match(pattern, t)) == True), '%s is not True' %t
for f in str_false:
    assert(bool(re.match(pattern, f)) == False), '%s is not False' %f
```

All numbers

```
pattern = '^-{0,1}\d*\.{0,1}\d+$'
str_true = ('1', '123', '1.234', '-123', '-123.0')
str_false = ('-abc')
for t in str_true:
    assert(bool(re.match(pattern, t)) == True), '%s is not True' %t
for f in str_false:
    assert(bool(re.match(pattern, f)) == False), '%s is not False' %f
```

6. Validating dates

Validates dates in *mm/dd/yyyy* format.

```
pattern = '^(0[1-9]|1[0-2])\/(0[1-9]|1\d|2\d|3[01])\/(19|20)\d{2}$'
str_true = ('01/08/2014', '12/30/2014', )
str_false = ('22/08/2014', '-123', '1/8/2014', '1/08/2014', '01/8/2014')
for t in str_true:
    assert(bool(re.match(pattern, t)) == True), '%s is not True' %t
for f in str_false:
    assert(bool(re.match(pattern, f)) == False), '%s is not False' %f
```

7. Checking for HTML tags

This regex is only recommended for "filtering" purposes and not a ultimate way to parse HTML.

```
pattern = r"""</?\w+((\s+\w+(\s*=\s*(?:".*?"|'.*?'|[^'">\s]+))?)+\s*|\s*)/?>"""

str_true = ('<a>', '<a href="something">', '</a>', '<img src>')

str_false = ('a>', '<a ', '< a >')

for t in str_true:

   assert(bool(re.match(pattern, t)) == True), '%s is not True' %t

for f in str_false:

   assert(bool(re.match(pattern, f)) == False), '%s is not False' %f
```

8. Checking for IP addresses

IPv4

```
pattern = r'^(?:(?:25[0-5]|2[0-4][0-9]|[01]?[0-9][0-9]?)\.){3}(?:25[0-5]|2[0-4][0-9]|[01]?[0-9][0-9]?)$'

str_true = ('172.16.254.1', '1.2.3.4', '01.102.103.104', )

str_false = ('17216.254.1', '1.2.3.4.5', '01 .102.103.104', )

for t in str_true:

   assert(bool(re.match(pattern, t)) == True), '%s is not True' %t

for f in str_false:

   assert(bool(re.match(pattern, f)) == False), '%s is not False' %f
```

IPv6

```
pattern = r'^\s*((([0-9A-Fa-f]{1,4}:){7}([0-9A-Fa-f]{1,4}|:))|(([0-9A-Fa-f]{1,4}:){6}(:[0-9A-Fa-f]{1,4}|((25[0-5]|2[0-4]\d|1\d\d|[1-9]?\d)(\.(25[0-5]|2[0-4]\d|1\d\d|[1-9]?\d)){3})|:))|(([0-9A-Fa-f]{1,4}:){5}(((:[0-9A-Fa-f]{1,4}){1,2})|:((25[0-5]|2[0-4]\d|1\d\d|[1-9]?\d)(\.(25[0-5]|2[0-4]\d|1\d\d|[1-9]?\d)){3})|:))|(([0-9A-Fa-f]{1,4}:){4}(((:[0-9A-Fa-f]{1,4}){1,3})|((:[0-9A-Fa-f]{1,4})?:((25[0-5]|2[0-4]\d|1\d\d|[1-9]?\d)(\.(25[0-5]|2[0-4]\d|1\d\d|[1-9]?\d)){3}))|:))|(([0-9A-Fa-f]{1,4}:){3}(((:[0-9A-Fa-f]{1,4}){1,4})|((:[0-9A-Fa-f]{1,4})
```

```
{0,2}:((25[0-5]|2[0-4]\d|1\d\d|[1-9]?\d)(\.(25[0-5]|2[0-4]\d|1\d\d|[1-9]?\d)){3}))|:))|(([0-9A-Fa-f]{1,4}:){2}(((:[0-9A-Fa-f]{1,4}){1,5})|((:[0-9A-Fa-f]{1,4}){0,3}:((25[0-5]|2[0-4]\d|1\d\d|[1-9]?\d)(\.(25[0-5]|2[0-4]\d|1\d\d|[1-9]?\d)){3}))|:))|(([0-9A-Fa-f]{1,4}:){1}(((:[0-9A-Fa-f]{1,4}){1,6})|((:[0-9A-Fa-f]{1,4}){0,4}:((25[0-5]|2[0-4]\d|1\d\d|[1-9]?\d)(\.(25[0-5]|2[0-4]\d|1\d\d|[1-9]?\d)){3}))|:))|(:(((:[0-9A-Fa-f]{1,4}){1,7})|((:[0-9A-Fa-f]{1,4}){0,5}:((25[0-5]|2[0-4]\d|1\d\d|[1-9]?\d)(\.(25[0-5]|2[0-4]\d|1\d\d|[1-9]?\d)){3}))|:)))(%.+)?\s*$'
str_true = ('2001:470:9b36:1::2',
        '2001:cdba:0000:0000:0000:0000:3257:9652',
        '2001:cdba:0:0:0:0:3257:9652',
        '2001:cdba::3257:9652', )
str_false = ('1200::AB00:1234::2552:7777:1313', # uses `::` twice
        '1200:0000:AB00:1234:O000:2552:7777:1313', ) # contains an
O instead of 0
for t in str_true:
   assert(bool(re.match(pattern, t)) == True), '%s is not True' %t
for f in str_false:
   assert(bool(re.match(pattern, f)) == False), '%s is not False' %f
```

Checking for MAC addresses

```
pattern = r'^(?i)([0-9A-F]{2}[:-]){5}([0-9A-F]{2})$'
str_true = ('94-AE-70-A0-66-83',
        '58-f8-1a-00-44-c8',
        '00:A0:C9:14:C8:29'
        , )
str_false = ('0:00:00:00:00:00',
        '94-AE-70-A0 -66-83', )
for t in str_true:
   assert(bool(re.match(pattern, t)) == True), '%s is not True' %t
for f in str_false:
   assert(bool(re.match(pattern, f)) == False), '%s is not False' %f
```

I believe that the above tips on using regular expression will profit many a readers.

Appendix 2

We will be introducing *pandas* library for data wrangling and analysis.

pandas is a Python package providing fast, flexible, and expressive data structures designed to work with *relational* or *labeled* data both. It is a fundamental high-level building block for doing practical, real world data analysis in **Python**.

pandas is well suited for:

Tabular data with heterogeneously-typed columns, as in an SQL table or Excel spreadsheet

Ordered and unordered (not necessarily fixed-frequency) time series data.

Arbitrary matrix data (homogeneously typed or heterogeneous) with row and column labels

Any other form of observational / statistical data sets. The data actually need not be labeled at all to be placed into a *pandas* data structure

We list some of the key features of the library below:

Easy handling of **missing data**

Size mutability: columns can be inserted and deleted from DataFrame and higher dimensional objects

Automatic and explicit **data alignment**: objects can be explicitly aligned to a set of labels, or the data can be aligned automatically

Powerful, flexible **group by functionality** to perform split-apply-combine operations on data sets

Intelligent label-based **slicing, fancy indexing, and subsetting** of large data sets

Intuitive **merging and joining** data sets

Flexible **reshaping and pivoting** of data sets

Hierarchical labeling of axes

Robust **IO tools** for loading data from flat files, Excel files, databases, and HDF5

Time series functionality: date range generation and frequency conversion, moving window statistics, moving window linear regressions, date shifting and lagging, etc.

Let us get started with using *pandas*.

```
from IPython.core.display import HTML
%matplotlib inline
import pandas as pd
import numpy as np
# Set some Pandas options
pd.set_option('html', False)
pd.set_option('max_columns', 30)
pd.set_option('max_rows', 20)
```

Pandas Data Structures

Series

A **Series** is a single vector of data (like a NumPy array) with an *index* that labels each element in the vector.

```
counts = pd.Series([632, 1638, 569, 115])
counts
0    632
1   1638
2    569
3    115
dtype: int64
```

If an *index* is not specified, a default sequence of integers is assigned as the index. A **NumPy** array comprises the values of the Series, while the *index* is a pandas **Index** object.

```
counts.values
array([ 632, 1638,  569,  115])
counts.index
Int64Index([0, 1, 2, 3], dtype='int64')
```

We can assign meaningful labels to the index, if they are available:

```
bacteria = pd.Series([632, 1638, 569, 115],
        index=['Firmicutes', 'Proteobacteria', 'Actinobacteria',
'Bacteroidetes'])

bacteria
Firmicutes      632
Proteobacteria  1638
Actinobacteria   569
Bacteroidetes    115
dtype: int64
```

These labels can be used to refer to the *values* in the **Series**.

```
bacteria['Actinobacteria']
569
bacteria[[name.endswith('bacteria') for name in bacteria.index]]
Proteobacteria  1638
Actinobacteria   569
dtype: int64
[name.endswith('bacteria') for name in bacteria.index]
[False, True, True, False]
```

Notice that the indexing operation preserved the association between the values and the corresponding indices. We can still use positional indexing if we wish.

```
bacteria[0]
632
```

We can give both the array of values and the index meaningful labels themselves:

```
bacteria.name = 'counts'
bacteria.index.name = 'phylum'
bacteria
phylum
Firmicutes      632
```

Proteobacteria 1638

Actinobacteria 569

Bacteroidetes 115

Name: counts, dtype: int64

NumPy's math functions and other operations can be applied to Series without losing the data structure.

np.log(bacteria)

phylumFirmicutes 6.448889

Proteobacteria 7.401231

Actinobacteria 6.343880

Bacteroidetes 4.744932

Name: counts, dtype: float64

We can also *filter* according to the values in the Series:

bacteria[bacteria>1000]

phylum

Proteobacteria 1638

Name: counts, dtype: int64

A **Series** can be thought of as an ordered *key-value* store. In fact, we can create one from a **dict**:

bacteria_dict = {'Firmicutes': 632, 'Proteobacteria': 1638, 'Actinobacteria': 569, 'Bacteroidetes': 115}

pd.Series(bacteria_dict)

Actinobacteria 569

Bacteroidetes 115

Firmicutes 632

Proteobacteria 1638

dtype: int64

Notice that the Series is created in key-sorted order.

If we pass a custom index to Series, it will select the corresponding values from the dict, and treat indices without corrsponding values as missing.

Pandas uses the **NaN** (not a number) type for *missing values.*

```
bacteria2 = pd.Series(bacteria_dict, index=['Cyanobacteria','Firmicutes','Proteobacteria','Actinobacteria'])
bacteria2
Cyanobacteria     NaN
Firmicutes        632
Proteobacteria   1638
Actinobacteria    569
dtype: float64
bacteria2.isnull()
Cyanobacteria     True
Firmicutes       False
Proteobacteria   False
Actinobacteria   False
dtype: bool
```

Critically, the labels are used to **align data** when used in operations with other Series objects:

```
bacteria + bacteria2
Actinobacteria   1138
Bacteroidetes     NaN
Cyanobacteria     NaN
Firmicutes       1264
Proteobacteria   3276
dtype: float64
```

Notice that the missing values were propogated by addition.

DataFrame

In practical situations, we want to be able to store, view and manipulate data that is *multivariate*, where for every index there are multiple fields or columns of data, often of varying data type.

A *DataFrame* is a tabular data structure, encapsulating multiple series like columns in a spreadsheet. Data are stored internally as a 2-dimensional object, but the DataFrame allows us to represent and manipulate higher-dimensional data.

```
data = pd.DataFrame({'value':[632, 1638, 569, 115, 433, 1130, 754, 555],
        'patient':[1, 1, 1, 1, 2, 2, 2, 2],
        'phylum':['Firmicutes', 'Proteobacteria', 'Actinobacteria',
    'Bacteroidetes', 'Firmicutes', 'Proteobacteria', 'Actinobacteria',
'Bacteroidetes']})
data
  patient        phylum  value
0       1    Firmicutes    632
1       1 Proteobacteria  1638
2       1 Actinobacteria   569
3       1  Bacteroidetes   115
4       2    Firmicutes    433
5       2 Proteobacteria  1130
6       2 Actinobacteria   754
7       2  Bacteroidetes   555
```

Notice the DataFrame is sorted by column name. We can change the order by indexing them in the order we desire:

```
data[['phylum','value','patient']]
         phylum  value  patient
0    Firmicutes    632        1
1 Proteobacteria  1638        1
2 Actinobacteria   569        1
3  Bacteroidetes   115        1
4    Firmicutes    433        2
5 Proteobacteria  1130        2
6 Actinobacteria   754        2
7  Bacteroidetes   555        2
```

A DataFrame has a second index, representing the columns:

data.columns

Index([u'patient', u'phylum', u'value'], dtype='object')

If we wish to access columns, we can do so either by *dict*-like indexing or by attribute:

data['value']

0 632

1 1638

2 569

3 115

4 433

5 1130

6 754

7 555

Name: value, dtype: int64

data.value

0 632

1 1638

2 569

3 115

4 433

5 1130

6 754

7

555

Name: value, dtype: int64

type(data.value)

pandas.core.series.Series

type(data[['value']])

pandas.core.frame.DataFrame

Notice this is different than with Series, where dict-like indexing retrieved a particular element (row). If we want access to a row in a DataFrame, we index its *ix* attribute.

```
data.ix[3]
patient              1
phylum   Bacteroidetes
value              115
Name: 3, dtype: object
```

Its important to note that the Series returned when a DataFrame is indexed it is merely a *view* on the DataFrame, and not a copy of the data itself. So you must be cautious when manipulating this data:

```
vals = data.value
vals
0    632
1   1638
2    569
3    115
4    433
5   1130
6    754
7    555
Name: value, dtype: object
vals[5] = 0
vals
0    632
1   1638
2    569
3    115
4    433
5      0
6    754
7    555
```

```
Name: value, dtype: object
data
  patient          phylum value
0       1      Firmicutes   632
1       1  Proteobacteria  1638
2       1  Actinobacteria   569
3       1   Bacteroidetes   115
4       2      Firmicutes   433
5       2  Proteobacteria     0
6       2  Actinobacteria   754
7       2   Bacteroidetes   555
vals = data.value.copy()
vals[5] = 1000
data
  patient          phylum value
0       1      Firmicutes   632
1       1  Proteobacteria  1638
2       1  Actinobacteria   569
3       1   Bacteroidetes   115
4       2      Firmicutes   433
5       2  Proteobacteria     0
6       2  Actinobacteria   754
7       2   Bacteroidetes   555
```

We can create or modify columns by assignment:

```
data.value[3] = 14
data
  patient          phylum value
0       1      Firmicutes   632
1       1  Proteobacteria  1638
2       1  Actinobacteria   569
```

```
3    1  Bacteroidetes    14
4    2     Firmicutes   433
5    2  Proteobacteria     0
6    2  Actinobacteria   754
7    2  Bacteroidetes   555
data['year'] = 2013
data
  patient         phylum value  year
0       1     Firmicutes   632  2013
1       1  Proteobacteria  1638  2013
2       1  Actinobacteria   569  2013
3       1   Bacteroidetes    14  2013
4       2     Firmicutes   433  2013
5       2  Proteobacteria     0  2013
6       2  Actinobacteria   754  2013
7       2   Bacteroidetes   555  2013
```

Specifying a Series as a new columns cause its values to be added according to the DataFrame's index:

```
treatment = pd.Series([0]*4 + [1]*2)
treatment
0  0
1  0
2  0
3  0
4  1
5  1
dtype: int64
data['treatment'] = treatment
data
  patient         phylum value  year  treatment
```

```
0    1     Firmicutes   632  2013       0
1    1  Proteobacteria  1638  2013       0
2    1  Actinobacteria   569  2013       0
3    1   Bacteroidetes    14  2013       0
4    2     Firmicutes   433  2013       1
5    2  Proteobacteria     0  2013       1
6    2  Actinobacteria   754  2013     NaN
7    2   Bacteroidetes   555  2013     NaN
```

Other Python data structures (ones without an index) need to be the same length as the DataFrame:

```
month = ['Jan', 'Feb', 'Mar', 'Apr']
data['month'] = month
---------------------------------------------------------------------------
ValueError                          Traceback (most recent call last)

<ipython-input-39-360d03fdde9a> in <module>()
      1 month = ['Jan', 'Feb', 'Mar', 'Apr']
----> 2 data['month'] = month
/usr/local/lib/python2.7/site-packages/pandas/core/frame.pyc in __setitem__(self, key, value)
   1899         else:
   1900             # set column
-> 1901             self._set_item(key, value)
   1902
   1903     def _setitem_slice(self, key, value):
/usr/local/lib/python2.7/site-packages/pandas/core/frame.pyc in _set_item(self, key, value)
   1981         is_existing = key in self.columns
   1982         self._ensure_valid_index(value)
-> 1983         value = self._sanitize_column(key, value)
   1984         NDFrame._set_item(self, key, value)
```

```
                                                        1985
/usr/local/lib/python2.7/site-packages/pandas/core/frame.pyc in _
sanitize_column(self, key, value)
  2031        elif isinstance(value, Index) or _is_sequence(value):
  2032            if len(value) != len(self.index):
-> 2033               raise ValueError('Length of values does not match
length of '
  2034                                'index')
  2035
ValueError: Length of values does not match length of index
```

The above value error can be circumvented by doing this:

```
data['month'] = ['Jan']*len(data)
data
```

	patient	phylum	value	year	treatment	month
0	1	Firmicutes	632	2013	0	Jan
1	1	Proteobacteria	1638	2013	0	Jan
2	1	Actinobacteria	569	2013	0	Jan
3	1	Bacteroidetes	14	2013	0	Jan
4	2	Firmicutes	433	2013	1	Jan
5	2	Proteobacteria	0	2013	1	Jan
6	2	Actinobacteria	754	2013	NaN	Jan
7	2	Bacteroidetes	555	2013	NaN	Jan

We can use **del** to remove columns, in the same way dict entries can be removed:

```
del data['month']
data
```

	patient	phylum	value	year	treatment
0	1	Firmicutes	632	2013	0
1	1	Proteobacteria	1638	2013	0
2	1	Actinobacteria	569	2013	0
3	1	Bacteroidetes	14	2013	0

```
4    2     Firmicutes  433  2013      1
5    2  Proteobacteria    0  2013      1
6    2  Actinobacteria  754  2013    NaN
7    2   Bacteroidetes  555  2013    NaN
```

We can extract the underlying data as a simple ndarray by accessing the values attribute:

```
data.values

array([[1, 'Firmicutes', 632, 2013, 0.0],

    [1, 'Proteobacteria', 1638, 2013, 0.0],

    [1, 'Actinobacteria', 569, 2013, 0.0],

    [1, 'Bacteroidetes', 14, 2013, 0.0],

    [2, 'Firmicutes', 433, 2013, 1.0],

    [2, 'Proteobacteria', 0, 2013, 1.0],

    [2, 'Actinobacteria', 754, 2013, nan],

    [2, 'Bacteroidetes', 555, 2013, nan]], dtype=object)
```

Notice that because of the mix of string and integer (and NaN) values, the dtype of the array is object. The dtype will automatically be chosen to be as general as needed to accomodate all the columns.

```
df = pd.DataFrame({'foo': [1,2,3], 'bar':[0.4, -1.0, 4.5]})
df.values
array([[ 0.4,  1. ],
    [-1. ,  2. ],
    [ 4.5,  3. ]])
```

Pandas uses a custom data structure to represent the indices of Series and DataFrames.

```
data.index
Int64Index([0, 1, 2, 3, 4, 5, 6, 7], dtype='int64')
```

Index objects are immutable:

```
data.index[0] = 15
---------------------------------------------------------------------------
TypeError                         Traceback (most recent call last)
<ipython-input-45-42a852cc9eac> in <module>()
----> 1 data.index[0] = 15


/usr/local/lib/python2.7/site-packages/pandas/core/base.pyc in _
disabled(self, *args, **kwargs)
  177     """This method will not function because object is immutable."""
   178      raise TypeError("'%s' does not support mutable operations." %
--> 179                     self.__class__)
   180
    181     __setitem__ = __setslice__ = __delitem__ = __delslice__ =
_disabled
TypeError: '<class 'pandas.core.index.Int64Index'>' does not support
mutable operations.
```

This is so that Index objects can be shared between data structures without fear that they will be changed.

```
bacteria2.index = bacteria.index
bacteria2
phylum
Firmicutes        NaN
Proteobacteria    632
Actinobacteria    1638
Bacteroidetes     569
dtype: float64
```

Importing data

A key, but often under-appreciated, step in data analysis is importing the data that we wish to analyze. Though it is easy to load basic data structures into Python using built-in tools or those provided by packages like NumPy, it is non-trivial to import structured data well, and to easily convert this input into a robust data structure. We do encounter with structured data, more often, in molecular biology. Let us see how to get started with importing data into pandas framework. Let us see, how we import in NumPy.

genes = np.loadtxt("genes.csv", delimiter=",", dtype=[('gene', '|S10'), ('value', '<f4')])

Pandas provides a convenient set of functions for importing tabular data in a number of formats directly into a DataFrame object. These functions include a slew of options to perform type inference, indexing, parsing, iterating and cleaning automatically as data are imported.

Let's start with some more bacteria data, stored in csv format. You can call it whatever you like.

```
Taxon,Patient,Tissue,Stool
Firmicutes,1,632,305
Firmicutes,2,136,4182
Firmicutes,3,1174,703
Firmicutes,4,408,3946
Firmicutes,5,831,8605
Firmicutes,6,693,50
Firmicutes,7,718,717
Firmicutes,8,173,33
Firmicutes,9,228,80
Firmicutes,10,162,3196
Firmicutes,11,372,32
Firmicutes,12,4255,4361
Firmicutes,13,107,1667
Firmicutes,14,96,223
Firmicutes,15,281,2377
Proteobacteria,1,1638,3886
Proteobacteria,2,2469,1821
Proteobacteria,3,839,661
```

Proteobacteria,4,4414,18
Proteobacteria,5,12044,83
Proteobacteria,6,2310,12
Proteobacteria,7,3053,547
Proteobacteria,8,395,2174
Proteobacteria,9,2651,767
Proteobacteria,10,1195,76
Proteobacteria,11,6857,795
Proteobacteria,12,483,666
Proteobacteria,13,2950,3994
Proteobacteria,14,1541,816
Proteobacteria,15,1307,53
Actinobacteria,1,569,648
Actinobacteria,2,1590,4
Actinobacteria,3,25,2
Actinobacteria,4,259,300
Actinobacteria,5,568,7
Actinobacteria,6,1102,9
Actinobacteria,7,678,377
Actinobacteria,8,260,58
Actinobacteria,9,424,233
Actinobacteria,10,548,21
Actinobacteria,11,201,83
Actinobacteria,12,42,75
Actinobacteria,13,109,59
Actinobacteria,14,51,183
Actinobacteria,15,310,204
Bacteroidetes,1,115,380
Bacteroidetes,2,67,0
Bacteroidetes,3,0,0

```
Bacteroidetes,4,85,5
Bacteroidetes,5,143,7
Bacteroidetes,6,678,2
Bacteroidetes,7,4829,209
Bacteroidetes,8,74,651
Bacteroidetes,9,169,254
Bacteroidetes,10,106,10
Bacteroidetes,11,73,381
Bacteroidetes,12,30,359
Bacteroidetes,13,51,51
Bacteroidetes,14,2473,2314
Bacteroidetes,15,102,33
Other,1,114,277
Other,2,195,18
Other,3,42,2
Other,4,316,43
Other,5,202,40
Other,6,116,0
Other,7,527,12
Other,8,357,11
Other,9,106,11
Other,10,67,14
Other,11,203,6
Other,12,392,6
Other,13,28,25
Other,14,12,22
Other,15,305,32
```

This table can be read into a DataFrame using *read_csv*:

```
mb = pd.read_csv("data/microbiome.csv")
```

mb

	Taxon	Patient	Tissue	Stool
0	Firmicutes	1	632	305
1	Firmicutes	2	136	4182
2	Firmicutes	3	1174	703
3	Firmicutes	4	408	3946
4	Firmicutes	5	831	8605
5	Firmicutes	6	693	50
6	Firmicutes	7	718	717
7	Firmicutes	8	173	33
8	Firmicutes	9	228	80
9	Firmicutes	10	162	3196
..	...	...	...	...
65	Other	6	116	0
66	Other	7	527	12
67	Other	8	357	11
68	Other	9	106	11
69	Other	10	67	14
70	Other	11	203	6
71	Other	12	392	6
72	Other	13	28	25
73	Other	14	12	22

```
74     Other     15    305    32

[75 rows x 4 columns]
```

Notice that *read_csv* automatically considered the first row in the file to be a header row.

We can override default behavior by customizing some the arguments, like header, names or index_col. We show one example here, reader is advised to check doccumentation for proper description and help on functions.

```
pd.read_csv("data/microbiome.csv", header=None).head()

          0        1      2      3

0     Taxon  Patient  Tissue  Stool

1 Firmicutes       1    632    305

2 Firmicutes       2    136   4182

3 Firmicutes       3   1174    703

4 Firmicutes       4    408   3946
```

read_csv is just a convenience function for read_table, since csv is such a common format:

mb = pd.read_table("data/microbiome.csv", sep=',')

The *sep* argument can be customized as needed to accomodate arbitrary separators. For example, we can use a regular expression to define a variable amount of whitespace, which is unfortunately very common in some data formats:

sep='\s+'

For a more useful index, we can specify the first two columns, which together provide a unique index to the data.

```
mb = pd.read_csv("data/microbiome.csv", index_col=['Taxon','Patient'])

mb.head()

                Tissue  Stool
Taxon      Patient
Firmicutes 1         632    305
           2         136   4182
           3        1174    703
           4         408   3946
           5         831   8605
```

This is called a *hierarchical* index, which we will revisit later.

If we have sections of data that we do not wish to import (for example, known bad data), we can populate the *skiprows* argument:

```
pd.read_csv("data/microbiome.csv", skiprows=[3,4,6]).head()

      Taxon  Patient  Tissue  Stool
0  Firmicutes     1     632    305
1  Firmicutes     2     136   4182
2  Firmicutes     5     831   8605
3  Firmicutes     7     718    717
4  Firmicutes     8     173     33
```

Conversely, if we only want to import a small number of rows from, say, a very large data file we can use nrows:

```
pd.read_csv("data/microbiome.csv", nrows=4)

      Taxon  Patient  Tissue  Stool
0  Firmicutes      1     632    305
1  Firmicutes      2     136   4182
2  Firmicutes      3    1174    703
3  Firmicutes      4     408   3946
```

Alternately, if we want to process our data in reasonable chunks, the *chunksize* argument will return an iterable object that can be employed in a data processing loop.

Most real-world data is *incomplete*, with values missing due to incomplete observation, data entry or transcription error, or other reasons. Pandas will automatically recognize and parse common missing data indicators, including NA and NULL.

```
Taxon,Patient,Tissue,Stool
Firmicutes,1,632,305
Firmicutes,2,136,4182
Firmicutes,3,,703
Firmicutes,4,408,3946
Firmicutes,5,831,8605
Firmicutes,6,693,50
Firmicutes,7,718,717
Firmicutes,8,173,33
Firmicutes,9,228,NA
Firmicutes,10,162,3196
Firmicutes,11,372,-99999
Firmicutes,12,4255,4361
Firmicutes,13,107,1667
Firmicutes,14,*,223
Firmicutes,15,281,2377
```

Proteobacteria,1,1638,3886
Proteobacteria,2,2469,1821
Proteobacteria,3,839,661
Proteobacteria,4,4414,18
Proteobacteria,5,12044,83
Proteobacteria,6,2310,12
Proteobacteria,7,3053,547
Proteobacteria,8,395,2174
Proteobacteria,9,2651,767
Proteobacteria,10,1195,76
Proteobacteria,11,6857,795
Proteobacteria,12,483,666
Proteobacteria,13,2950,3994
Proteobacteria,14,1541,816
Proteobacteria,15,1307,53
Actinobacteria,1,569,648
Actinobacteria,2,1590,4
Actinobacteria,3,25,2
Actinobacteria,4,259,300
Actinobacteria,5,568,7
Actinobacteria,6,1102,9
Actinobacteria,7,678,377
Actinobacteria,8,260,58
Actinobacteria,9,424,233
Actinobacteria,10,548,21
Actinobacteria,11,201,83
Actinobacteria,12,42,75
Actinobacteria,13,109,59
Actinobacteria,14,51,183
Actinobacteria,15,310,204

Bacteroidetes,1,115,380
Bacteroidetes,2,67,0
Bacteroidetes,3,0,0
Bacteroidetes,4,85,5
Bacteroidetes,5,143,7
Bacteroidetes,6,678,2
Bacteroidetes,7,4829,209
Bacteroidetes,8,74,651
Bacteroidetes,9,169,254
Bacteroidetes,10,106,10
Bacteroidetes,11,73,381
Bacteroidetes,12,30,359
Bacteroidetes,13,51,51
Bacteroidetes,14,2473,2314
Bacteroidetes,15,102,33
Other,1,114,277
Other,2,195,18
Other,3,42,2
Other,4,316,43
Other,5,202,40
Other,6,116,0
Other,7,527,12
Other,8,357,11
Other,9,106,11
Other,10,67,14
Other,11,203,6
Other,12,392,6
Other,13,28,25
Other,14,12,22
Other,15,305,32

```
pd.read_csv("data/microbiome_missing.csv").head(20)
          Taxon  Patient Tissue  Stool
0     Firmicutes       1    632    305
1     Firmicutes       2    136   4182
2     Firmicutes       3    NaN    703
3     Firmicutes       4    408   3946
4     Firmicutes       5    831   8605
5     Firmicutes       6    693     50
6     Firmicutes       7    718    717
7     Firmicutes       8    173     33
8     Firmicutes       9    228    NaN
9     Firmicutes      10    162   3196
10    Firmicutes      11    372 -99999
11    Firmicutes      12   4255   4361
12    Firmicutes      13    107   1667
13    Firmicutes      14      ?    223
14    Firmicutes      15    281   2377
15 Proteobacteria      1   1638   3886
16 Proteobacteria      2   2469   1821
17 Proteobacteria      3    839    661
18 Proteobacteria      4   4414     18
19 Proteobacteria      5  12044     83
```

Above, Pandas recognized NA and an empty field as missing data. We can specify additional symbols with the *na_values* argument. In the example above, we have a missing data with * and -99999 values. This can be handled while importing as follows:

```
pd.read_csv("data/microbiome_missing.csv",                na_values=['*',
-99999]).head(20)
```

	Taxon	Patient	Tissue	Stool
0	Firmicutes	1	632	305
1	Firmicutes	2	136	4182
2	Firmicutes	3	NaN	703
3	Firmicutes	4	408	3946
4	Firmicutes	5	831	8605
5	Firmicutes	6	693	50
6	Firmicutes	7	718	717
7	Firmicutes	8	173	33
8	Firmicutes	9	228	NaN
9	Firmicutes	10	162	3196
10	Firmicutes	11	372	NaN
11	Firmicutes	12	4255	4361
12	Firmicutes	13	107	1667
13	Firmicutes	14	NaN	223
14	Firmicutes	15	281	2377
15	Proteobacteria	1	1638	3886
16	Proteobacteria	2	2469	1821
17	Proteobacteria	3	839	661
18	Proteobacteria	4	4414	18
19	Proteobacteria	5	12044	83

These can be specified on a *column-wise* basis using an appropriate **dict** as the argument for *na_values*.

This small introduction to pandas library should rncourage the reader to delve into the complete functionalities of the package. The reader is encouraged to go through the following links to learn about pandas.

http://pandas.pydata.org/

http://pandas.pydata.org/pandas-docs/stable/

INDEX

L

M

N

O

P

R

S

T

U

W

X